J. Varma

MICROSCOPY HANDBOOKS 15

Dictionary of Light Microscopy

ROYAL MICROSCOPICAL SOCIETY
MICROSCOPY HANDBOOKS

---

01 Savile Bradbury: *An introduction to the optical microscope*

02 Dawn Chescoe and Peter Goodhew: *The operation of the transmission electron microscope*

03 Peter Goodhew: *Specimen preparation for transmission electron microscopy of materials*

04 J. James and J. Tas: *Histochemical protein staining methods*

05 A. J. Morgan: *X-ray microanalysis in electron microscopy for biologists*

06 Olga Bayliss High: *Lipid histochemistry*

07 Derek Hemsley: *The light microscopy of synthetic polymers*

08 S. K. Chapman: *Maintaining and monitoring the transmission electron microscope*

10 J. S. Ploem and H. J. Tanke: *Introduction to fluorescence microscopy*

11 Julia Polak and Susan Van Noorden: *An introduction to immunocytochemistry: current techniques and problems* (revised edition)

12 Andrew Briggs: *An introduction to scanning acoustic microscopy*

13 D. J. Thomson and Savile Bradbury: *An introduction to photomicrography*

14 J. D. Bancroft and N. M. Hand: *Enzyme histochemistry*

15 S. Bradbury, P. J. Evennett, H. Haselmann, and H. Piller: *Dictionary of light microscopy*

In preparation

09 P. C. Robinson: *Qualitative polarized light microscopy*

MICROSCOPY HANDBOOKS 15

Dictionary of Light Microscopy

Compiled by the Nomenclature Committee of the RMS

S. Bradbury
Department of Human Anatomy, Oxford

P.J. Evennett
Department of Pure and Applied Biology, Leeds

H. Haselmann
Institut für Wissenschaftliche Mikroskopie, Tübingen

H. Piller
Aalen, West Germany

Oxford University Press · Royal Microscopical Society · 1989

Oxford University Press, Walton Street, Oxford OX2 6DP
Oxford New York Toronto
Delhi Bombay Calcutta Madras Karachi
Petaling Jaya Singapore Hong Kong Tokyo
Nairobi Dar es Salaam Cape Town
Melbourne Auckland
and associated companies in
Berlin Ibadan

Royal Microscopical Society,
37/38 St. Clements,
Oxford OX4 1AJ

Published in the United States
by Oxford University Press, New York

British Library Cataloguing in Publication Data
RMS dictionary of light microscopy.
1. Microscopy
I. Royal Microscopical Society.
Nomenclature Committee
502'.8'2
ISBN 0-19-856413-9

Library of Congress Cataloging in Publication Data
RMS dictionary of light microscopy/compiled by the Nomenclature Committee of the RMS,
S. Bradbury . . . [et al.].
p. cm.—(Microscopy handbooks; 15)
1. Microscope and microscopy—Dictionaries. I. Bradbury, Savile. II. Royal Microscopical Society (Great Britain). Nomenclature Committee. III. Series.
QH203.R57 1988 502'.8'2—dc19 88-22712
ISBN 0-19-856413-9 (pbk.)

Set by Pentacor Ltd., High Wycombe, Bucks
Printed in Great Britain
at the University Printing House, Oxford
by David Stanford
Printer to the University

Introduction

This dictionary is the result of several years' hard work by the Nomenclature Committee of the Royal Microscopical Society. Its recommendations are made after careful consideration of current usage, international standards (where appropriate), and the origin and meanings of the words involved. We recognize that many familiar objects are given unfamiliar names and that many familiar terms may be absent; some terms have been adapted from the German where no precise equivalent exists in English. Some terms have been marked as those whose use we would wish to discourage, because they are incorrect, they refer to obsolete equipment, or are inconsistent with other preferred terms. We hope that our recommended terms will be used as much as possible in manufacturers' publications, on teaching courses, and in scientific papers. We have included many cross references but have deliberately not been entirely consistent in positioning the definitions. This is in order to place them where we feel that they are most likely to be sought.

Although a few terms in quantitative microscopy and photometry have been included, these topics have not been exhaustively covered and we hope that these (and others) will form the subjects of more specialized publications in due course. This first attempt at defining microscopical terms is not to be regarded as final: we aim to update it in the light of readers' comments and corrections. It will be helpful if any such comments are accompanied by a suggested new or replacement definition or entry for the dictionary and sent to the office of the Society.

We are grateful to Dr L. Bernstein of the Institut d'Optique, Orsay, France for providing the French equivalent terms listed in Appendices II–IV.

S.B. P.J.E.
H.H. H.P.

1988

SI units of length used in microscopy

Unit	Multiplier	Symbol
Metre	1	m
Millimetre	10^{-3}	mm
Micrometre	10^{-6}	μm
Nanometre	10^{-9}	nm
Picometre	10^{-12}	pm

The ångstrom unit (Å), equal to 10^{-10} m or 10^{-1} nm, is not an SI unit of length, but may be encountered in older literature.

parallel to the *optical axis* to *focus* at a different distance from those in the plane at right angles to it. (2) Differences in *refractive power* in different meridional sections of the *eye*, caused either by irregular curvature of the *cornea* (corneal astigmatism) or by irregularities in the *refractive power* of the lens (lens astigmatism).

autofluorescence (primary fluorescence) The inherent *fluorescence* capability of a substance or tissue component.

autoradiography A technique for demonstrating the location of a radioactively labelled tracer substance in a tissue or organ. The radioactivity produces developable silver grains in a photographic film or emulsion placed in contact with the specimen and kept in the dark for the period of exposure to the radioactivity.

auxiliary microscope *See* **telescope, auxiliary**

auxiliary telescope *See* **telescope, auxiliary**

axial chromatic aberration *See* **aberration, chromatic, axial**

axial figure [pol.] A special type of *conoscopic interference figure* in the form of a symmetrical pattern of concentric rings or *lemniscates* intersected by a dark cross or hyperbolae. It arises when an optically *uniaxial object* is observed along its *crystal optical axis* or when an optically *biaxial object* is observed along a *bisectrix*.

axial illumination *See* **illumination, axial**

axial magnification *See* **magnification, axial**

axial ray *See* **ray, axial**

axis, crystal optical [pol.] The direction of *light* propagation in an optically *anisotropic* medium for which the substance behaves as if it were *isotropic*. The crystal optical axis may be referred to simply as the optical axis when the usage is clear from the context.

axis, optic The *optical axis* of the *eye*. (*See also* **visual axis (of the eye)**

axis, optical A straight line joining the centres of curvature of *lens* surfaces.

axis, visual (of the eye) *See* **visual axis (of the eye)**

azimuth angle The angle between a direction lying in the reference plane, or the projection of a direction into this plane (the 'trace'), and a *reference direction* lying in this plane. In microscopy the azimuth angle is described by reference to the *object plane* (see Fig. 3) and is important in describing the direction of non-symmetrical illumination or imaging methods.

B

Babinet, Jacques (1794–1872) A French physicist interested in optical instrumentation. The inventor of a compensator which bears his name.

back focal plane *See focal plane, back*

balsam, Canada A resin obtained from the balsam fir tree (*Abies balsamea*) used as the basis of a microscopical *mounting medium*.

bandwidth, half-peak-height The width of the band of *wavelengths* lying between the points on the transmission curve of a *filter*, at which the *transmission* is one half of its maximum value (at the *peak wavelength*).

barrel distortion *See* **distortion, barrel**

barrier filter *See* **filter, barrier**

base, microscope That part of the microscope *stand* which rests on the work table and to which the other parts of the instrument are attached. In modern instruments the base may be box-shaped and contain parts of the illuminating system. (see Fig. 1)

beam *See* **beam, light**

beam, incident Any *beam*, the path of which intercepts a surface.

beam, light A group of parallel or approximately unidirectional *rays*.

beam splitter A means whereby a *beam* may be divided into two or more separate beams.

Becke, Friedrich J. K. (1855–1931) An Austrian mineralogist and petrologist who contributed to the methodology of petrological research. (*See* **Becke line**)

Becke line A bright line (due to *refraction* and/or *diffraction*) formed in the *image* at the boundary between media of different *optical pathlengths*. It moves in the direction of the longer optical path when the distance between the *objective* and the *object* is increased.

Note: This phenomenon is used to recognize relative differences in refractive index of two adjacent media, e.g. a particle and the surrounding mounting medium. When the refractive indices are matched the Becke line disappears.

Beer-Lambert law The basic quantitative expression of the *absorbance* of electromagnetic *radiation*. The absorbance (A) of a solution is equal to the product of the absorptivity (a constant ϵ), the concentration of solute (c) and the thickness (t) of the layer of solution through which the radiation passes. Thus

$$A = \epsilon t c.$$

Berek, Max (1886–1949) A German physicist and mathematician (associated with the firm of E. Leitz) who designed many optical instruments, in particular for polarized light microscopy.

Bertrand diaphragm *See* **diaphragm, Bertrand**

Bertrand lens (Amici–Bertrand lens) *See* **lens, Bertrand**

biaxial, optically [pol.] Having two *crystal optical axes*.

binocular Using both eyes.

binocular head *See* **tube, binocular**

binocular microscope *See* **microscope, binocular**

binocular tube *See* **tube, binocular**

bireflectance (pol.] The maximum difference in *reflectance* due to *bireflection*.

bireflection [pol.] The variation of colour and/or *brightness* of *optically anisotropic* reflecting objects if viewed with *plane-polarized light* as the orientation of the *vibration direction* is changed.

Note: The description of bireflection is related to the principal vibration directions or, exceptionally, to crystallographic or morphological features.

birefraction [pol.] *See* **double refraction**

birefringence [pol.] The quantitative expression of the maximum difference in *refractive index* due to *double refraction*, (symbol Δn).

birefringence, sign of [pol.] The designation of *birefringence* as positive (+) or negative (−) according to the following convention. For optically *uniaxial* media the birefringence is defined as the difference between the 'extraordinary' *principal refractive index* and the 'ordinary' one. It is expressed as $\Delta n = n_e - n_o$. If $n_e > n_o$, Δn is positive; if $n_e < n_o$ then Δn is negative. For optically *biaxial* media the principal refractive indices (x, y, and z) are involved in the following way:

$$\Delta n = (n_z - n_y) - (n_y - n_x).$$

If $(n_z - n_y) > (n_y - n_x)$, Δn is positive.
If $(n_z - n_y) < (n_y - n_x)$, Δn is negative.

Note: The sign of birefringence applies only to transparent media. The subscripts e, o, x, y, z may be replaced by the Greek letters ϵ, ω, a, β, γ respectively.

bisectrix [pol.] The line which bisects an *optic axial angle*.

black body A body which completely absorbs, and hence does not reflect, any incident *radiation*.

Note: In practice, such a body is inevitably imperfect; a fair approximation is achieved from a hollow space surrounded by a wall which is impermeable to radiation but which has a small opening in it. Radiation incident onto the wall is almost completely absorbed on the inside of the wall.

black-body radiation *See* **radiation, black-body**

black-body radiator *See* **radiator, black-body**

blooming *See* **coating of lens surfaces**

body tube *See* **tube, body**

body-tube-locating surfaces *See* **locating surface (or flange)**

Brewster, David (1781–1868) A Scottish author and tutor. He was interested and carried out researches in optical mineralogy and astronomy as well as in polarized light and microscopy.

bright-field *See* **microscopy, bright-field**

bright-field illumination *See* **illumination, bright-field**

bright-field microscopy *See* **microscopy, bright-field**

brightness (1) A qualitative expression of the strength of the sensation of light. (2) A synonym for *luminance*.

broad-band-pass filter *See* **filter, broad-band-pass**

Brownian movement *See* **movement, Brownian**

bulb The envelope (usually glass or fused silica) of a *filament lamp*. A term commonly used to describe the *lamp* itself.

bundle, ray All those *rays* bounded by a *diaphragm*.

bundle, ray, converging A bundle of *rays* which converges or appears to converge towards a single point.

bundle, ray, diverging A bundle of *rays* which originates or appears to originate from a single point.

bundle, ray, parallel A bundle of *rays* which neither converge nor diverge.

C

camera head *See* **head, camera**

camera lucida *See* **drawing apparatus**

camera tube *See* **head, camera**

candela [photom.] The *photometric* SI base unit. It is defined as the *luminous intensity* (in the direction perpendicular to the surface) of an area of $1/(6 \times 10^5)\text{m}^2$ of a *black-body radiator*, at the temperature of the solidification of platinum (2042 K) at a pressure of 1 standard atmosphere.

Note: Since 1978 the 'standard atmosphere' has been replaced by the 'bar' (1 atmosphere = 1.01325 bar).

cardinal elements *See* **elements, cardinal**

cardinal distances *See* **distances, cardinal**

cardinal planes *See* **planes, cardinal**

cardinal points *See* **points, cardinal**

cardioid condenser *See* **condenser, cardioid**

catadioptric An adjective used to describe an optical system which operates both by *reflection* and *refraction*.

catoptric An adjective used to describe an optical system which operates by *reflection*.

cement A material of specified optical properties used between optical surfaces to hold them together permanently.

central stop *See* **stop, central**

central wavelength *See* **wavelength, central**

centring nosepiece *See* **nosepiece, centring**

chamber, cold A special *slide* consisting of an enclosed space whose temperature can be lowered for maintaining a *specimen* below room temperature whilst it is observed.

chamber, counting A special *slide* which, when covered with a suitable *cover glass*, forms a cavity of a predetermined height. The base of the chamber is engraved with squares of defined size so that the number of particles contained within a defined volume of liquid introduced into the chamber may be counted.

chamber, culture A special *slide* consisting of an enclosed space arranged to contain a nutrient medium, intended for observing living cells whilst they grow.

chamber, warm A special *slide* consisting of an enclosed space whose temperature can be raised for the maintenance of a *specimen* above room temperature whilst it is observed.

chromatic (1) Coloured. (2) Pertaining to or consisting of *colour*. (3) Relating to *wavelength*.

chromatic aberration *See* **aberration, chromatic**

chromatic colour *See* **colour, chromatic**

chromatic difference of magnification *See* **aberration, chromatic, lateral**

chromaticity [photom.] The *colour* quality of *light* as defined by its *chromaticity co-ordinates* or, alternatively, by its *dominant* (or complementary) *wavelength* and its *purity* taken together.

chromaticity co-ordinates [photom.] The ratios of any one of the *tristimulus values* of a *colour* sample to the sum of its three tristimulus values. Symbolized by the letters x, y, z.

chromaticity diagram [photom.] A graph on which the *chromaticity co-ordinates* x and y are plotted (see Fig. 10.)

circle of least confusion The smallest diameter *image* spot formed from a point *object* when *spherical aberration* and *astigmatism* are present.

circularly polarized light *See* **light, circularly polarized**

coarse adjustment A *focusing mechanism* designed to make large and rapid alterations in the distance along the *optical axis* between the *object* and the *objective*.

coating of lens surfaces (blooming) The deposit of one or more thin dielectric and/or metallic layers on a surface of an optical element (e.g. *lens*, mirror, *prism*, or *filter*) for the purpose of decreasing or increasing *reflection*.

code, colour, of objectives *See* **colour code of objectives**

coherence The correspondence of different *wavetrains* with respect to several distinct properties (defined as the *coherence condition*) which enable the wavetrains to *interfere*.

Note: Coherent radiation can be produced only by a very small source (a point source), by a very small part of a larger source, or by a laser.

coherence, degree of The proportion of *coherent radiation* to total radiation in a mixture of *coherent* and *incoherent radiation*.

coherence, partial Description of *radiation* which consists of both *coherent* and *incoherent* components. Expressed quantitatively as the *degree of coherence*.

coherence condition The relationship between two or more *wavetrains* (emanating from a common *source*) which must exist in order for *interference* to occur. The waves must act at the same time on the same point in space and (over a sufficiently long period in time—the observation period) the phase relations between the waves must remain constant. In addition, the *wavelengths* must be equal, and the waves must vibrate in the same plane. At extended sources the divergence angle between the direction of propagation of wavetrains which may interfere must fulfil the condition

$$n \sin u \leqslant \lambda/2b$$

where n is the *refractive index* of the medium through which the *radiation* travels, u is the maximum angle between the directions of propagation of wavetrains which may interfere, b is the width of the source, and λ is the wavelength of the radiation.

coherent Satisfying the *coherence condition*.

cold chamber *See* **chamber, cold**

collar, correction *See* **correction collar**

collecting power of a lens The property of a *converging lens* or concave mirror to render a *parallel ray bundle* convergent or to reduce the divergence of a *diverging ray bundle*.

collector A *lens* which serves to project a suitably sized *image* of the *source* into a given plane (e.g. in *Köhler illumination*, into the plane of the *aperture* of the *condenser*). Sometimes known as a 'lamp collector'.

collector diaphragm *See* **diaphragm, collector**

collimate To make *rays* originating from one point parallel to one another.

collimation The making parallel of *rays*.

collimator A device to *collimate rays*, usually a *converging lens* or a concave mirror.

colorimetry [photom.] The measurement and objective description of parameters characterizing *colours* by means of standard procedures and terms.

colour A basic visual sensation, photometrically characterized by *chromaticity* and *luminance factor* or *luminous transmittance*.

colour, achromatic [photom., physiol.] The *colours* white, black, and grey.

colour, chromatic [photom., physiol.] *Colours* other than white, black, and grey.

colour, interference A mixed *colour* resulting from extinction or partial extinction caused by *interference* of one or several parts of a *spectrum*.

colour code of objectives Systems of *colour marking*. A coloured band applied to the *mount* of an *objective* indicates a range of *magnifying power*. The colours may be assigned from black to white, in spectral order, to denote increasing *magnifying power*. Thus

1–1.25	black
1.6–2	grey
2.5–3.2	brown
4–5	red
6.3–8	orange
10–12.5	yellow
16–20	bright green
25–32	dark green
40–50	light blue
63–80	dark blue
100,125,160	white.

A second colour band may be used to indicate the kind of *immersion liquid*: black indicates oil, white indicates water, orange indicates glycerol, and red indicates other liquids.
The colour of the engraving denotes the imaging mode for which the objective is designed, e.g. red signifies *polarized light*, green signifies *phase contract*.

colour-conversion filter *See* **filter, colour-conversion**

colour filter *See* **filter, colour**

colour marking of objectives Coloured rings and/or engraving applied to an *objective* to denote its properties in accordance with a *colour code*.

colour saturation *See* **saturation of colour**

colour temperature *See* **temperature, colour**

colour values, tristimulus [photom.] The amounts of each of the three primary *colours* which must be combined to form an objective colour match with the sample, according to the rules of the Commission International de l'Eclairage.

coma (1) An *aberration* in which the *image* of an off-axis point *object* is deformed so that the image is shaped like a comet. (2) The aberration of *lenses* and mirrors which is responsible for the above defect.

comparison microscope *See* **microscope, comparison**

compensating eyepiece *See* **eyepiece, compensating**

compensator [pol.] A *retardation plate* of variable *optical pathlength difference* used to measure the optical pathlength differences within an *object*. Many types exist, often designated by the name of their originator e.g. Babinet, Berek, Senarmont.

complementary dominant wavelength *See* **wavelength, complementary dominant**

complex radiation *See* **radiation, complex**

compound microscope *See* **microscope, compound**

compressarium A form of *slide* designed to restrain a moving organism between two plates of glass whose spacing is variable and through which the organism may be observed.

condenser A part of the *illuminating system* of the *microscope* which consists of one or more *lenses* (or mirrors) and their *mounts*, usually containing a *diaphragm*, and is designed to collect, control, and concentrate *radiation*. In *bright-field microscopy* by *reflected-light illumination* the *objective* serves as its own condenser.

condenser diaphragm *See* **diaphragm, condenser**

condenser, Abbe A *condenser* of simple design introduced by Ernst Abbe, in which there is only limited *correction* for *spherical aberration* and none for *chromatic aberration*.

condenser, achromatic–aplanatic A *condenser* in which *spherical* and *chromatic aberration* have been partially corrected permitting a high *numerical aperture* and the use of oil *immersion* if desired.

condenser, cardioid A *darkground condenser* for *transmitted light*, in which the *correction* for *spherical aberration* and *coma* is calculated for a reflecting surface with the shape of a cardioid of revolution. In practice, the correction is achieved by using a suitable zone of a spherical surface which differs imperceptibly in its corrective effect from a true cardioid surface.

condenser, darkground (darkfield) A *condenser* designed for *darkground microscopy*. For *transmitted-light microscopy* this condenser is a separate component; for *reflected-light-microscopy* it is generally within the objective mount surrounding the imaging system of the *objective*.

condenser, lamp *See* **collector**

condenser, pancratic A *condenser* containing a variable (*pancratic*) *lens* which allows the size of the *illuminated field* at the *object* to be varied while the *illuminated field diaphragm* remains of constant size. The size of the *aperture* varies inversely with that of the illuminated field at the object but the product of both sizes remains a constant.

condenser, paraboloid A *darkground condenser* for *transmitted-light microscopy*, now almost obsolete, in which the spherical *corrections* are achieved by means of a reflecting surface with the shape of a paraboloid of revolution.

condenser, phase-contrast A *condenser* designed for *phase-contrast microscopy*. It forms on the *phase plate* in the *back focal plane* of the *objective* a suitably sized image of a *diaphragm* (generally annular) positioned in the *front focal plane* of the condenser.

condenser, substage A *condenser* designed for the *substage of a microscope*.

condenser, swing-out top lens A type of *condenser* designed so that its top *lens* can conveniently be removed from the optical path by operating a lever, thus increasing the condenser's *focal length* in order to increase the area of the *illuminated field* and decrease the *numerical aperture* for use with *objectives* of low *magnifying power*.

condition, coherence *See* **coherence condition**

cone, aplanatic That part of the illuminating cone, formed by a *condenser*, which is free from *spherical aberration* and *coma*.

cones (of the retina) [physiol.] One of the two types of sensory cell in the *retina*. They serve for perception in *daylight (photopic) vision* and in *twilight (mesopic) vision*. *Colour* sensation is mediated by the cones only.

confocal imaging mode *See* **imaging mode, confocal**

confusion, circle of least *See* **circle of least confusion**

conjugate Linked together in accordance with the rules of *geometrical optics*.

conjugate planes Those planes perpendicular to the *optical axis* at the *conjugate points*.

conjugate points Those points in both *object space* and *image space* which are imaged one on to the other.

conjugate portions of a ray *See* **rays, conjugate portions of**

conoscopic (interference) figure [pol.] A pattern consisting of *isogyres* and/or *isochromatic curves* formed in the *back focal plane* of the *objective* when an optically *anisotropic object* is placed between *crossed polars*, or, exceptionally, *parallel polars*.

conoscopic observation (or conoscopy) [pol.] Observation of the *conoscopic (interference) figure* by means of a pin-hole *diaphragm* or *auxiliary telescope* in place of the *eyepiece*, or by means of a *Bertrand lens*.

continuous spectrum *See* **spectrum, continuous**

contrast Generally taken to mean *physiological contrast*; when *photometric contrast* is meant, this should be stated explicitly.

contrast, differential interference *See* **differential interference contrast**

contrast, interference *See* **interference contrast**

contrast, marginal [physiol.] An apparent increase in *contrast* perceived in a narrow zone on each side of the boundary between two areas of different *brightness* and/or *chromaticity*.

contrast, modulation *See* **microscopy, relief contrast**

contrast, photometric Ratios of *intensities* or of differences between intensities in two areas in an optical plane.

Note: Contrast between areas 1 and 2 may be expressed in one of three ways.

$(X_2 - X_1)/(X_2 + X_1)$, X_2/X_1, or $(X_2 - X_1)/X_2$

where X is any *radiometric* or *photometric* quantity and the subscript 2 indicates the greater quantity. Photometric contrast may also be called *intensity* contrast (sometimes, incorrectly, referred to as *amplitude* contrast).

contrast, physiological [physiol.] The phenomenon caused by the stimulus distribution in the *visual organ* when viewing areas of differing *brightness* and/or *chromaticity*. This phenomenon consists of a difference in sensation which does not correspond to the temporal or local distribution of stimuli on the *retina*. If the difference is perceived mainly as *brightness*, it is called brightness contrast; if it relates mainly to *chromaticity*, it is called colour contrast. Contrast is perceived differently depending on whether the stimuli are presented simultaneously in adjacent fields (simultaneous contrast) or successively (successive contrast). If the contrast is of the simultaneous type, then *marginal contrast* is important in the recognition of structures.

contrast, relief *See* **relief contrast**

contrast filter *See* **filter, contrast**

convergence (1) (In *geometrical optics*) the property of a *bundle of rays* coming together in the direction of *light* propagation. (2) [physiol.] The inward turning (adduction) of the *visual axes* of both *eyes*. The angle between the visual axes is the convergence angle.

convergence angle *See* **convergence**

convergence point *See* **point, convergence**

converging lens *See* **lens, converging**

converging ray bundle *See* **bundle, ray, converging**

cooling stage *See* **stage, cooling**

co-ordinates, chromaticity *See* **chromaticity co-ordinates**

cornea [physiol.] The anterior, strongly curved, transparent part of the exterior wall of the eyeball through which the image-forming *rays* pass. The iris of the *eye* and its *pupil* are visible through the cornea. The curvature of the front surface of the cornea provides the major part of the total *refractive power* of the eye.

correction The process whereby the *aberrations* of an optical system are minimized.

correction class The type of *correction* of an optical system (*achromatic, plan-achromatic*, etc.).

correction collar A mechanism provided on some *objectives* in order to adapt their correction for *spherical aberration* to compensate for deviations from correct *optical pathlength* in the *medium* or media between the objective and the *object*.

correction for image distance The calculation of a *microscope objective* to optimize its *corrections* for a given standardized *image distance*. (*See also* **objective, image distance correction of**)

counting chamber *See* **chamber, counting**

counting eyepiece *See* **eyepiece, counting**

cover glass (cover slip) A rectangular or circular piece of thin glass used to cover a *microscopical preparation*. For calculation and *correction* of the *objective* it is regarded as part of the objective and thus its thickness, *refractive index* and *dispersive power* must be adapted to the demands of the objective. These parameters are standardized in ISO Standard No. 8255/1.

critical, angle The *angle of incidence* of a *ray* passing from a medium of higher to one of lower *refractive index* at which the ray just fails to emerge and is totally internally reflected. The sine of this angle is the ratio of the lower refractive index to the higher refractive index.

critical illumination *See* **illumination, source-focused**

cross, extinction *See* **extinction cross**

cross hairs *See* **cross lines**

cross lines A *graticule* with a cross-shaped pattern used to indicate the centre of the *image field* and/or lateral reference directions, e.g. *vibration direction* of *polars*.

cross wires *See* **cross lines**

crossed polars *See* **polars, crossed**

crown glass *See* **glass, crown**

crystal optical axis *See* **axis, crystal, optical**

crystalline lens *See* **lens of the eye**

culture chamber *See* **chamber, culture**

curvature of image field *See* **image field, curvature of**

curve, dispersion *See* **dispersion curve**

curve, extinction *See* **extinction curve**

curve, spectral luminosity *See* **spectral luminosity curve**

D

dark adaptation [physiol.] The process of adjustment of the *visual organ* from higher to lower *luminances*.

darkfield *See* **microscopy, darkground**

darkground condenser *See* **condenser, darkground**

darkground illumination *See* **illumination, darkground**

darkground microscopy *See* **microscopy, darkground**

darkground stop A central opaque disc usually used in the *front focal plane* of a *condenser* to occlude all the *direct light* which would fall within the *aperture* of the *objective*.

daylight vision *See* **vision, daylight**

definition The perception of the degree of *resolution* in an *image*.

degree of coherence *See* **coherence, degree of**

degree of polarization *See* **polarization, degree of**

density, optical Equivalent to *absorbance*.

depth of field (depth of sharpness in object space) The axial depth of the space on both sides of the *object plane* within which the *object* can be moved without detectable loss of sharpness in the *image*, and within which features of the object appear acceptably sharp in the image while the position of the *image plane* is maintained.

depth of focus (depth of sharpness in image space) The axial depth of the space on both sides of the *image plane* within which the *image* appears acceptably sharp while the positions of the *object plane* and of the *objective* are maintained.

diagonal position *See* **position, diagonal**

diagram, chromaticity *See* **chromataticity diagram**

diaphragm A mechanical limitation of an opening normal to the *optical axis* which restricts the cross-sectional area of the *light path* at a defined place in an optical system. It may be fixed or variable in size, shape (although usually circular) and position.

diaphragm, aperture A *diaphragm* in the plane of any *aperture* of an optical system which delimits its *pupil*.

diaphragm, Bertrand [pol.] A *field diaphragm*, placed after a *Bertrand lens*, which restricts the *field* from which a *conoscopic figure* is formed.

diaphragm, collector A *diaphragm* which controls the opening of a *collector* and which is usually made *conjugate* with *field planes*.

diaphragm, condenser A *diaphragm* which controls the effective size and shape of the *condenser aperture*. A synonym for the *illuminating aperture diaphragm* in *transmitted-light illumination*.

diaphragm, field A *diaphragm* in the *object plane* or in any plane *conjugate* with it. Field diaphragms are usually fitted just after the *lamp collector* and in the *eyepiece*.

diaphragm, fixed A *diaphragm* of fixed size and shape.

diaphragm, illuminated field A *diaphragm* whose *image* delimits the *illuminated field* in the *object*. Commonly known as the *field diaphragm* or field stop.

diaphragm, illuminating aperture An *aperture diaphragm* which delimits the *pupil* of an *illuminating system*. For *transmitted light* this is usually incorporated in the *front focal plane* of the *condenser*; in *reflected-light microscopes* it is found in the *illuminator* in a plane conjugate with the *back focal plane* of the *objective*.

diaphragm, iris A *diaphragm* bounded by multiple leaves, usually metal, arranged so as to provide an opening of variable size, which is adjustable by means of a control.

diaphragm, photometric [photom.] A *diaphragm* which delimits the *photometric field*.

diaphragm, photomicrographic field A *diaphragm* or *mask* which delimits a photomicrographic field. It is the *field diaphragm* of the camera.

diaphragm, visual field A *diaphragm* which delimits a *visual field*. It is usually contained within the *eyepiece*.

dichroic (dichromatic) mirror A special type of *interference filter* used as an essential part of a fluorescence *microscope* using *epi-illumination*. It is designed to reflect selectively the shorter wavelength exciting radiation and transmit the longer wavelength *fluorescence*. A similar device is often used as a *lamp* reflector, in order to transmit the longer wavelength heat (*infrared*) radiation while reflecting the light.

dichroism [pol.] A case of *pleochroism* occurring in optically *uniaxial* media, in which the *absorption* depends on the *polarization state*.

differential interference *See* **interference, differential**

differential interference contrast *Contrast* formed by the phenomenon of *differential interference*.

diffracted light *Light* which has undergone *diffraction* at the *object*. It gives rise to the first-order, second-order, etc. components of the *diffraction pattern*.

diffraction Deviation of the direction of propagation of *light* or other wave motion when the wavefront passes the edge of an obstacle.

diffraction, Fraunhofer *Diffraction* of parallel *light* (in which the incident wavefronts are planar). The resulting pattern is received on a *screen* infinitely distant from the diffracting *object* or in the *back focal plane* of a *lens* following the diffracting object.

diffraction, Fresnel *Diffraction* of *light* diverging from a *source* at a finite distance from an *object*. A pattern is received on a *screen* at a finite distance from the diffracting object.

diffraction apparatus, Abbe *See* **Abbe diffraction apparatus**

diffraction disc The *image* of a primary or secondary *point source of light* which, due to *diffraction* at a circular *lens aperture*, takes the form of a bright disc surrounded by a sequence of concentric dark and light rings. Sometimes known as the 'Airy disc'.

diffraction grating *See* **grating, diffraction**

diffraction limit of resolving power A fundamental limitation imposed upon the *resolving power* of a system by the phenomenon of *diffraction* alone, and not by *aberration*.

diffraction pattern (or diffraction image) A distribution of *intensities* varying with direction in a regular manner and resulting from *interference* between portions of the diffracted *radiation* having differing phase relationships.

diffraction pattern (or image), primary The *Fraunhofer diffraction* pattern of a *microscopic* object formed in the *back focal plane* of the *objective*.

Note: Since Abbe called this pattern the 'primary interference image', confusion may arise with the 'primary (real) image' as defined in this dictionary. To avoid this, the word 'pattern' should be used instead of the word 'image' when referring to the primary diffraction pattern.

diffuse reflection *See* **reflection, diffuse**

diffusion Change of the spatial distribution of a *beam* when it is deviated in all directions by a surface or by a *medium*, without change of the proportions of the *monochromatic* components of which the *radiation* is composed.

dimensions, optical fitting (of the microscope) Mechanical or opto-mechanical distances measured from *reference planes* on which the calculation of *microscope lenses* and the design of *microscopes* are based, and which facilitate the interchange of certain components. There are two categories of dimension: those which are standardized internationally and others which are taken as internal standards by individual manufacturers. (See Fig. 2 and Table 2.)

dimensions, optical fitting, reference plane for A surface of a *microscope* component (e.g. a *locating surface* or *flange*) or a plane in the *light path* of the microscope used as a limit for one of the *optical fitting dimensions*. (See Fig. 2.)

DIN An abbreviation for Deutsches Institut für Normung (formerly Deutsche Industrie Norm), the German authority for standardization. The same abbreviation is used in photography to denote a method for expressing film speed in logarithmic units.

dioptre A unit of *refractive power* expressed as the reciprocal of the *focal length* of a *lens* in metres.

dioptric An adjective used to describe optical arrangements or optical elements, indicating that they operate by *refraction*, i.e. using *lenses*.

dipping cone (or cap) An attachment which fits onto the front of a dry *objective* to allow it to operate beneath the surface of a liquid. Frequently used for the observation of living aquatic organisms.

direct light *Light* which enters the *objective* after undergoing no change in direction of propagation on passing through the *object field* (*transmitted light*), or on *specular reflection* at a flat surface in the object field orientated normally to the direction of propagation of the light (*reflected light*). It gives rise to the zero-order component of the diffraction image of the *illuminating aperture diaphragm* formed in the *back focal plane* of the objective.

direction, extinction *See* **extinction direction**

direction, vibration (1) Direction of the electric vector of electromagnetic *radiation*, i.e. the direction in the *vibration plane* normal to the direction of wave propagation. (2) [pol.] Used in *polarized-light microscopy* to describe the vibration direction of *wavetrains* selected by a *polar* or of wavetrains passing through an *optically anisotropic* medium.

directions, references [pol.] Cartesian coordinates to which directions of orientation and/or *vibration* are referred. The basic direction is the x-axis, designated as positive when it runs from left to right (or west to east) as seen by an observer looking towards the *light source*. The *vibration direction* of the *polarizer* is arranged to coincide with this axis. The y-axis is perpendicular to the x-axis in the *field* and runs from south to north. The z-axis coincides with the *optical axis* running in the direction of light propagation. Anticlockwise rotation (as seen from the observer's viewpoint looking towards the light source) in the x–y plane about the z-axis is designated positive. Now standardized in ISO Standard No. 8576.

directions, vibration, principal [pol.] Mutually perpendicular *vibration directions* of an *optically anisotropic* medium parallel to the symmetry axes of the *optical indicatrix*.

Note: For an optically uniaxial medium the vibration direction parallel to the axis of revolution of the appropriate indicatrix is designated as the extraordinary vibration direction (symbol e or ϵ) and the vibration direction perpendicular to this axis is designated as the ordinary (symbol o or ω). For an optically biaxial medium the longest axis of the appropriate indicatrix is symbolized by z or γ, the vibration direction parallel to the shortest axis by x or α and the vibration direction parallel to the third axis by y or β.

disc, diffraction *See* **diffraction disc**

discharge lamp *See* **lamp, discharge**

disparity [physiol.] The distance between the *image* point in one *eye* and the retinal point in the same eye which corresponds with the image point of the same object point in the other eye. A distinction must be made between transverse and vertical disparity.

dispersion (1) The change in *phase velocity* of a *wavegroup* as a function of its *wavelength* (or frequency) when passing from one *medium* to another which causes a separation of the *monochromatic* components of a *complex radiation*. (2) The variation in *refractive index* of a medium which gives rise to the above phenomenon. (3) The quantity characterizing this property; it may have a special name, e.g. the *Abbe number* or the *dispersive power* of the medium.

dispersion curve A graph of *refractive index* as a function of *wavelength* or a related parameter.

dispersion staining microscopy *See* **microscopy, dispersion staining**

dispersive power The reciprocal of the *Abbe number*.

Note: The wavelengths to which the refractive indices refer may be slightly different from those taken as reference for the Abbe number.

dissecting microscope A low-power *microscope* of long *free working distance* used for dissecting; nowadays generally a *stereomicroscope*.

distance, accommodation [physiol.] The distance of an *accommodation* point from the *nodal point* of the *eye* on the object side.

distance, far-point [physiol.] The distance between the object-side *principal point* of the *eye* and the *farpoint*.

distance, free working *See* **objective, free working distance of**

distance, image The distance (a') along the *optical axis* of a *lens* between the image-side *principal plane* (H′) and the *image plane* (O′). The image distance is one of the *cardinal distances*. (*See* Fig. 4.)

Note: The image distance is not to be confused with the object to primary image distance, which is especially important in microscopy.

distance, interpupillary [physiol.] The distance in millimetres between the centres of the *pupils* of a person's *eyes* when viewing with parallel *fixation lines*. *Binocular microscopes* and *binocular tubes* are provided with an adjustment to allow for the variable interpupillary distances of different people.

distance, intersection The distance (symbol s) between the *object point* O and the object-side *lens vertex* (object-side intersection distance) or the distance s' between the *image point* O′ and the image-side lens vertex (image-side intersection distance). The intersection distance is one of the *cardinal distances*. In practice this distance is important with regard to the *free working distance* of a lens, e.g. *objective* or *condenser*, because the position of the *principal planes* is not normally known. (See Fig. 4.)

distance, minimum resolvable *See* **resolvable distance, minimum**

distance, near-point [physiol.] The distance between the object-side *principal point* of the *eye* and the *near point* (*See also* **reference viewing distance)**

distance, object The distance (a) along the *optical axis* of a *lens* between the object-side *principal plane* (H) and the *object plane* (O). (*see* Fig. 4.) The object distance is one of the *cardinal distances*.

distance, object to primary image The distance in air between the *object plane* and the *primary image plane*. It is the fundamental *optical fitting dimension* used in *microscope* design and commonly has a value of either 195 mm or infinity. The latter is a hypothetical value applied to microscopes designed for *infinity-corrected objectives*. (See Fig. 2 and Table 2.)

distance, objective to primary image The distance in air between the *objective locating surface* (of the *nosepiece*) and the *primary plane*. It is one of the *optical fitting dimensions* and commonly has a value of either 150 mm or infinity. The latter is a hypothetical value applied to *microscopes* designed for *infinity-corrected objectives*. (See Fig. 2 and Table 2.)

distance, optical *See* **pathlength, optical**

distance, reference viewing *See* **reference viewing distance**

distance, resolved *See* **resolved distance**

distances, cardinal In *geometrical optics* the distances along the optical *axis* between the *cardinal points* of a *lens*. They are used when considering the imaging function and for the representation of the *ray path* in diagrams. The cardinal distances are (Fig. 4): the *focal lengths* (f on the object side, f' on the image side); the *object distance* a; the *image distance* a'; the *intersection distances* (s at the object side and s' on the image side); and the *hiatus* (i).

distortion The *aberration* in which *lateral magnification* varies with distance from the *optical axis* in the *image field*.

distortion, barrel A difference in *lateral magnification* between the central and peripheral areas of an *image* such that the lateral magnification is less at the periphery. A square *object* in the centre of the *field* thus appears barrel shaped (i.e. with convex sides).

distortion, pincushion A difference in *lateral magnification* between the central and the peripheral areas of an *image* such that the lateral magnification is greater towards the periphery. A square *object* in the centre of the *field* thus appears pincushion shaped (i.e. with concave sides).

distribution temperature *See* **temperature, distribution**

divergence (1) The property of *rays* in a *diverging ray bundle*. (2) [physiol.] The outward turning (abduction) of the *visual axes* of the *eyes*. The angle between the visual axes is the divergence angle.

divergence angle *See* **divergence**

diverging lens *See* **lens, diverging**

diverging ray bundle *See* **bundle, ray, diverging**

dominance, ocular [physiol.] The dominance of one *eye*, which occurs frequently in normal *binocular* vision.

dominant wavelength *See* **wavelength, dominant**

double-beam interference *See* **interference, double-beam**

double-focus interference *See* **interference, double-focus**

double refraction An effect of *optical anisotropy*; the division of electromagnetic waves into *plane-polarized* components having mutually perpendicular *vibration directions* and being propagated with different velocities. Double refraction may be due to crystal properties (intrinsic *double refraction*), to the presence of orientated particles in a *medium* of different *refractive index* (form double refraction) or to strain in parts of the optical system or *object* (strain double refraction).

doublet A combination of two *lenses* of differing optical properties used to produce an *image* in which *aberrations* (especially spherical and chromatic) are reduced.

drawing apparatus A *microscope* accessory intended to facilitate the production of accurate representations of the *image* by means of drawing. It allows the microscope image and the plane of the drawing paper to be observed simultaneously.

drawing prism A type of *drawing apparatus* consisting of a special *prism* mounted above the *eyepiece*.

drawtube A sliding *viewing tube* (not fitted to most current microscopes) which enables the *mechanical tubelength* of the *microscope* to be adjusted in order to optimize the *correction* for *spherical aberration*, e.g. when the spherical correction is affected by the use of a *coverglass* of incorrect thickness. It may also be used in special cases to make small changes in the *lateral magnification* of the *primary image*.

dry objective *See* **objective, dry**

E

efficacy (of the eye), maximum spectral luminous [photom., physiol.] The maximum *luminous flux* produced by unit *radiant flux* and observed under specified conditions. The quantity depends on the *wavelength*. It occurs at 555 nm for *daylight vision* (where it is 673 *lumen* per watt) and at 507 nm for *night vision* (where it is 1725 lumen per watt). It is a calculation constant (symbolized by K_{max}) *for converting photometric quantities* into *radiometric* ones or vice versa. (See Table 1.)

efficiency, luminous [photom.] The ratio of the *luminous flux* of *complex radiation* to the corresponding *radiant flux*.

Note: Not to be confused with 'overall luminous efficiency', which is the ratio of the luminous flux to the total input power.

efficiency (of the eye), spectral luminous [photom., physiol.] The ratio of the *luminous flux* produced at a specified *wavelength* by unit *radiant flux* to the *maximum luminous efficiency*; it is represented by the *luminosity curve*, has a maximum value of 1 (or 100%) and serves to convert *photometric quantities* into *radiometric quantities* or vice versa for wavelengths outside those at which the *maximum luminous efficacy* occurs. There is a difference between the relative spectral luminous efficiency for *daylight vision* and for *night vision*.

electronic flash *See* **flash, electronic**

elements, cardinal A collective term for *cardinal distances*, *points* and *planes*. (See Figs. 4 and 5.)

elliptically polarized light *See* **light, elliptically polarized**

emission Release of *radiant energy*.

emmetropia [physiol.] Normal vision in which the *far point* is at infinity.

empty magnification *See* **magnification, empty**

energy, radiant Energy emitted, transferred or received in the form of *radiation* (see Table 1).

England finder A type of *object finder*.

entoptic phenomena [physiol.] Visual sensations arising from *objects* located within the *eye* itself, or spurious sensations arising within those parts of the brain which contribute to the visual process. In *microscopy* they generally result from inhomogeneities of the optical media of the eye itself and are more distinct when the diameter of the *exit pupil* of the *microscope* is very small. Sometimes referred to as '*muscae volitantes*'.

entrance pupil *See* **pupil**

entrance pupil of the eye *See* **pupil, entrance, of the eye**

envelope The glass or fused silica covering of a filament or discharge *lamp*.

epi-illumination *See* **illumination, epi-**

excitation The input of energy to matter leading to the *emission* of *radiation*.

exciter filter *See* **filter, exciter**

exit pupil *See* **pupil**

exit pupil of the microscope *See* **microscope, exit pupil of**

exitance, luminous (at a point of a surface) [photom.] The *luminous flux* leaving a unit area of surface. (See Table 1.)

exitance, luminous (at a point on a surface) [photom.] The *radiant flux* leaving a unit area of surface. (See Table 1.)

exposure The total quantity of *light* allowed to fall upon a photosensitive emulsion. Measured in *lux* per second.

exposure, light [photom.] Surface density of the *quantity of light* received or the product of *illuminance* and its duration. (See Table 1.)

exposure, radiant (at a point on a surface) [photom.] Surface density of *radiant energy* received or the product of *irradiance* and its duration. (See Table 1.)

exposure meter A device for determining the required *exposure* for photographic materials.

extended source *See* **source, extended**

external-diaphragm eyepiece *See* **eyepiece, external-diaphragm**

extinction [pol.] The condition in which an *optically anisotropic object* appears dark when observed between crossed *polars*, provided that at least one of the following conditions is met: the *light* passes along a *crystal optical axis*; the object is oriented so that its *vibration* directions are parallel to those of the crossed polars; the object is illuminated with *monochromatic light* and the *optical pathlength difference* for waves in the two vibration directions is equal to, or an integral multiple of, the *wavelength* of that light.

extinction, oblique [pol.] *Extinction* which occurs if one of the *vibration directions* of an *optically anisotropic object* forms a certain *azimuth angle* with a straight reference line, e.g. a pronounced crystallographic or morphological feature.

extinction, straight [pol.] *Extinction* which occurs if one of the *vibration directions* of an *optically anisotropic object* is parallel to a straight reference line, e.g. a pronounced crystallographic or morphological feature.

extinction, symmetrical [pol.] *Extinction* which occurs if one of the *vibration directions* of an *optically anisotropic object* bisects an *azimuth angle* between two straight reference lines, e.g. pronounced crystallographic or morphological features.

extinction angle [pol.] The *azimuth angle* at which *oblique extinction* occurs.

extinction cross [pol.] A pattern in the form of a dark cross formed in the *back focal plane* of the *objective* when its *aperture* is fully illuminated and the *polars* are crossed but no *object* is present (*transmitted light*), or when an *optically isotropic* and specularly reflecting surface such as a mirror serves as the object (*reflected light*).

Note: The pattern arises by the rotation of the plane of vibration of light emanating from the polarizer at curved lens surfaces and/or by axially symmetrical strain in lenses. It is more distinct with higher numerical apertures because of the stronger curvatures of the lenses.

extinction curve [pol.] The stereographic plot of all *extinction directions* arising from the various spatial directions of *light* propagation through an *optically anisotropic object* at constant orientation with respect to *crossed polars*.

extinction direction [pol.] An imaginary direction which coincides with one of the *vibration directions* of an *optically aniso-*

tropic object if this is in the *extinction position*.

extinction factor [pol.] A measure of the influence on the perfection of *extinction* with *crossed polars* of other optical components (such as *lenses* or glass plates). It is expressed as the ratio $X_{||}/_{+}$ where $X_{||}$ and X_{+} are the *intensities* measured when the whole unit (consisting of the *polars* and the test component between them) is illuminated with parallel (||) and crossed (+) polars.

extinction position [pol.] The state of orientation of a *double refracting* medium so that it shows *extinction* between *crossed polars*.

extraordinary ray *See* **ray, extraordinary**

eye Part of the *visual organ* in which an *image* of the external world is produced and transformed into nerve impulses; these, after transmission to the brain, are processed in order to give rise to visual sensations. In *microscopy* the eye is an integral part of the optical system.

eye, lens of *See* **lens of the eye**

eyelens The *lens* or group of lenses of an *eyepiece* nearest to the observer's *eye*.

eyepiece (or ocular) A *lens system* which is responsible for the *angular magnification* of the final *virtual image* formed by it at infinity from the *primary image*. This *image* is converted into a *real image* by the observer's *eye* or other *converging lens* system.

eyepiece, comparison A device for linking two *microscopes* for use as a *comparison microscope*.

eyepiece, compensating An *eyepiece* designed to correct *chromatic difference of magnification* in the *primary image*. It may also correct other off-axial defects, e.g. *astigmatism*.

eyepiece, counting A *focusable eyepiece* in which a *graticule* is fitted to define areas within which *objects* may be counted.

eyepiece, demonstration An *eyepiece*, often fitted with a pointer, to which a *beam splitter* and second viewing tube can be attached to permit simultaneous observation by two people.

eyepiece, external-diaphragm An *eyepiece* in which the *field diaphragm* is located in front of the *lenses*.

eyepiece, flat-field An *eyepiece* computed to reduce the *curvature of field* in the final *image*.

eyepiece, focusable An *eyepiece* with a mechanism for focusing a (interchangeable) *graticule* or *diaphragm* mounted within it and coinciding with the *primary image*.

eyepiece, goniometer A *focusable eyepiece* fitted with a *graticule* for measuring angles in the *microscope image*.

eyepiece, graticule A *focusable eyepiece* designed to carry a *graticule*.

eyepiece, high-eyepoint An *eyepiece* computed so that the *exit pupil of the microscope* is further from the *eyelens* than in a normal eyepiece of the same *focal length*. It is intended to facilitate use of the microscope by wearers of spectacles and/or for special applications.

* **eyepiece, Homal** A trade name for a special type of *negative eyepiece* used in order to produce a *real image*. It was developed for use in place of a conventional eyepiece as a projection system for the purpose of flattening the field in *photomicrography*.

eyepiece, Huygenian (1) A term originally used for an *eyepiece* consisting of two planoconvex lenses (the *field lens* and the *eyelens*) mounted with their convex sides facing the *objective*. The separation and *focal length* of these lenses provides achromatism. (2) A term now commonly used for any *internal-diaphragm eyepiece*.

eyepiece, image shearing A type of *micrometer eyepiece* in which two *images* of adjustable separation (and usually in complementary colours) are formed by an optical system. Dimensions are derived from the amount of adjustment required for the images to move from complete overlap to the point at which they are just separate.

eyepiece, integrating A *focusable eyepiece* equipped with a *graticule* in the form of a grating carrying lines, crosses, or points of known separation, used to facilitate *modal analysis* and evaluation of stereological parameters.

eyepiece, internal-diaphragm An eyepiece in which the *field diaphragm* is located between the *field lens* and the *eyelens* in the *front focal plane* of the eyelens.

eyepiece, Kellner An improved type of *Ramsden eyepiece*, in which the *eyelens* is an *achromatic doublet*.

eyepiece, locating flange of The flange on the *eyepiece* which locates it at a given level (that of the *eyepiece-locating surface of the viewing tube*). It is one of the reference planes for the *parfocalizing distance of the eyepiece* (see Fig. 2).

eyepiece, micrometer A *focusable eyepiece* used for measuring. In its most common form a measuring *graticule* is fitted in the *primary image plane*. It must be calibrated against a *stage micrometer*.

eyepiece, micrometer-screw A type of *micrometer eyepiece* in which reference marks in the *primary image plane* may be adjusted by means of a micrometer screw; the resultant indicated displacement is used to derive dimensions.

* **eyepiece, negative** Not a true *eyepiece* in the original meaning of the word, but a *diverging lens* acting as a *projection lens* and used mainly in *photomicrography*. It is not usable for visual observation, since its *exit pupil* is located within the system and is thus not accessible to the *pupil* of the observer's *eye*. An old and incorrect term for an *internal diaphragm eyepiece*. (*See also* **eyepiece, Homal**)

eyepiece, orthoscopic *See* **eyepiece, Kellner**

eyepiece, parfocalizing distance of *See* **parfocalizing distance (of eyepiece)**

eyepiece, pointer An *eyepiece* containing a movable pointer in its *primary image plane* serving to indicate *objects* of interest in the *field of view*.

* **eyepiece, positive** An old and incorrect term for an *external diaphragm eyepiece*.

eyepiece, projection Not a true *eyepiece* in the original meaning of the term, but a *lens* designed to project an *image* at a finite distance.

eyepiece, Ramsden The original type of *external-diaphragm eyepiece* in which the separation and *focal lengths* of the *lenses* provided achromatism.

eyepiece, slotted [pol.] An *eyepiece* containing a slot into which a *retardation plate* or other device may be inserted.

eyepiece, widefield An *eyepiece* specially computed to provide a *field of view* greater than that of a normal eyepiece of the same *magnifying power*.

eyepiece graticule *See* **graticule, eyepiece-**

eyepiece-locating surface (of viewing tube) The surface at the upper end of the *viewing tube* which sets the level of the *locating flange of the eyepiece*. It is one of the reference planes which determines the *mechanical tubelength*. (See Fig. 2.)

eyepoint *See* **microscope, exit pupil of**

eyepoint height The distance measured along the *optical axis* from the last surface of the *eyepiece* to the *eyepoint*. Its value may be affected by optical systems which are inserted between *objective* and eyepiece.

F

factor, extinction *See* **extinction factor**

factor, luminance *See* **luminance factor**

far-point distance *See* **distance, far-point**

far point of the eye [physiol.] The point at which the naked eye is focused when it is unaccommodated (i.e. with its lowest *refractive power*). This distance from the *eye* depends on possible refractive errors of the eye.

farsightedness *See* **hypermetropia**

fibre optic A bundle of very fine, optically insulated glass or plastic fibres used to conduct *light*. They may serve for *illumination* or for *image* transmission.

fibre optic illuminator *See* **illuminator, fibre optic**

field An area in the *object plane* or any other plane conjugate with it. The term may be qualified by its location (e.g. *object field*, *image field*) or its function (e.g. *illuminated field*, *photometric field*).

field, depth of *see* **depth of field**

field, illuminated That part of the *object field* which receives *illumination*.

field, image Any *field* in which an *image* of the *object* is formed.

field, object That part of the *object plane* which can be surveyed at any one time. Its *image* is delimited by the *field diaphragm*.

field, photometric That part of the *image field* from which a photometric measurement is made at any one time.

field, surrounding (1) That part of the *field of view* surrounding the object of vision. (2) That part of the field of view surrounding any *object*.

field, visual *See* **field of view**

field diaphragm *See* **diaphragm, field**

field lens A *lens* positioned in or close to a *field plane* in order to adapt the *exit pupil* of the preceding lenses to the *entrance pupil* of subsequent lenses. This suppresses vignetting in the *image* and, more generally, provides homogeneous *illumination* of the *field* to which it relates.

Note: The term is, unfortunately, often used without qualification to describe the field lens of the eyepiece.

field microscope *See* **microscope, portable**

field of view That part of the *image field* which is imaged on the observer's *retina*, and hence can be surveyed at any one time.

field-of-view number A number which specifies the *field of view* of an *eyepiece*. It is the actual diameter in millimetres of the *field diaphragm* in an *external-diaphragm eyepiece* or the apparent diameter of the *virtual image* of the field diaphragm in an *internal-diaphragm eyepiece*.

Note: The field-of-view number is now one of the standard markings of the eyepiece and may be used to calculate the diameter of the object field.

field plane *See* **plane, field**

figure, conoscopic *See* **conoscopic (interference) figure**

figure, polarization [pol.] The qualitative estimation of deformations of the *extinction cross* during rotation about the *optical axis* of an *optically aniotropic* specularly reflecting surface due to the influence of anisotropism.

Note: The phenomenon is used to identify ore minerals.

filament The emissive element of an incandescent electric *lamp*; usually made in the form of a coil of tungsten wire heated by the passage of an electric current. In low-voltage lamps this coil is often flattened to decrease the depth of the coil in the direction of the *optical axis* and to increase the *luminance*.

filament lamp *See* **lamp, filament**

filar micrometer *See* **eyepiece, micrometer-screw**

filter An optical device designed to control selectively a given range or all of the *wavelengths*, *colour temperature*, *vibration direction*, and/or *intensity* of the *radiation* which it transmits or reflects.

filter, barrier A *filter* used in *fluorescence microscopy* which is designed to prevent the passage towards the *image* of those *wavelengths* of *light* used for excitation but to allow the light produced by *fluorescence* of the *specimen* to pass.

filter, broad-band-pass (or broad-band) A *filter* which allows the passage of *radiation* with a broad *wavelength band* (greater than about 50 nm) around a given *central wavelength*.

Note: This concept of a 'broad' band is arbitrary.

filter, colour A *filter* which allows the passage of *light* of selected *colour* (*chromaticity*), or *wavelength* characteristics.

filter, colour-conversion (or conversion) A *filter* used to change the *colour temperature* of *light* received from a *source*.

filter, contrast A *filter* used to increase the *contrast* in an *image* between features of an *object* or between the object and the background.

filter, exciter A *filter* used in *fluorescence microscopy* designed (ideally) to pass only those *wavelengths* which excite *fluorescence*.

filter, heat (or heat protection) A *filter* designed to prevent the passage of *radiation* in the *infra-red* or near infra-red ranges which may cause damage to the *object* and/or optical elements. (*See also* **dichroic mirror**)

filter, interference A *filter* designed to transmit or reflect selectively a limited part of the *spectrum* by *multiple-beam interference*.

filter, long-wave-pass (or long-pass) A *filter* designed to allow the passage of *radiation* of *wavelengths* longer than a given limit.

filter, narrow-band-pass (or narrow-band) A *filter* (often of the *interference* type) which allows the passage of *radiation* only within a very narrow *wavelength band* around a given *central wavelength*.

Note: The concept of a 'narrow' band is arbitrary.

filter, neutral-density (or neutral) A *filter* designed to reduce as equally as possible the *intensity* of *radiation* across the whole visible *spectrum*.

filter, polarizing A *filter* acting as a *polar*, by total or partial *absorption* of *light* vibrating in certain directions.

filter, short-wave-pass (or short-pass) A *filter* designed to allow the passage of *radiation* of *wavelengths* shorter than a given limit.

filter tray An open, flat, and low-rimmed receptacle used for supporting a *filter* or filters, often attached to the *substage*.

finder *See* **object finder**

fine adjustment A *focusing mechanism* designed to make small and precise alterations in the relative positions along the *optical axis* between the object and the *objective*. The precision of positioning which it provides should be better than the *depth of field* of the objective.

first-order red *See* **red, first-order**

first-order red plate *See* **plate, first-order red**

fixation [physiol.] The alignment of the *eye* to look at an object point. In a normal eye this alignment occurs in such a way that the fixed *object* point is imaged in the centre of the *fovea*.

fixation line *See* **line, fixation**

fixation point *See* **point, fixation**

fixed diaphragm *See* **diaphragm, fixed**

flash, electronic A discharge tube and its associated components, designed to emit an intense pulse of *light* of short duration. In *microscopy* it is generally used for the *photomicrography* of moving objects.

flat-field eyepiece *See* **eyepiece, flat-field**

flat-field objective *See* **objective, flat-field**

flint glass *See* **glass, flint**

fluorescence A form of *photoluminescence* which persists only for a very short time (usually of the order of less than 10^{-8} s) after the cessation of the *excitation*.

fluorescence, auto- *See* **fluorescence, primary**

fluorescence microscope *See* **microscope, fluorescence**

fluorescence microscopy *See* **microscopy, fluorescence**

fluorescence, primary *Fluorescence* exhibited by virtue of the inherent properties of an *object*.

fluorescence, secondary *Fluorescence* exhibited by an *object* after treatment with a *fluorochrome*.

fluorimetry *Photometry* in which the measurement of *fluorescence* phenomena is involved.

fluorimetry, microscope *Microscope photometry* in which the measurement of *fluorescence* phenomena is involved.

fluorite Crystalline calcium fluoride (CaF_2). Because of its low *dispersion* this material is used as an additional *lens* material for the *correction* of *chromatic aberration* in some *microscope objectives* of the higher *correction classes*, e.g. *apochromats* and *semi-apochromats*.

* **fluorite objective** *See* **semi-apochromat**

fluorochrome A substance used to impart *fluorescence* to structures within a *specimen* for subsequent examination by *fluorescence microscopy*.

fluorophore A chemical group which confers *fluorescence*.

flux, luminous [photom.] The time rate of flow of a quantity of *light* emitted, transferred, or received. Its unit is the *lumen*. (See Table 1.)

Note: This is a basic photometric quantity.

flux, radiant [photom.] The power emitted, transferred, or received in the form of *radiation* or the time rate of flow of *radiant* (electromagnetic) *energy*. Its unit is the watt. (See Table 1.)

Note: This is a basic radiometric radiometric.

focal depth *See* **focus, depth of**

focal length The distance (f or f') measured along the *optical axis* from the *principal plane* to the appropriate *focal plane* (see Fig. 4.) It is one of the *cardinal distances*.

focal plane (1) A surface connecting all the points at which *bundles* of parallel *rays* entering an ideal converging *lens* cross on the other side of the lens, and thus containing a *focal point*. (2) A surface at right angles to the *optical axis* of a lens (or mirror) in which the *image* of an *object* lying at infinity is formed. It is one of the *cardinal planes*.

focal plane, back The *focal plane* of a *lens* which lies behind it when viewed in the direction of passage of *light*.

focal plane, front The *focal plane* of a *lens* which lies in front of it when viewed in the direction of the passage of *light*.

focal point The point of intersection (F or F′) of the *focal plane* with the *optical axis*, and where *rays* entering an ideal converging *lens* parallel to the optical axis cross the optical axis on the other side of the lens. It is one of the *cardinal points*. (See Fig. 4.)

focus (1) A synonym for the *focal point* of a *lens*. (2) The state of sharpest imaging. (3) A *point* in the *object plane* at which those *rays* intersect which, after *refraction* and/or *reflection* in an optical system also intersect in the *image plane* to give rise to a sharp *image* of the *conjugate point*. (4) The act of bringing the optical system into focus, i.e. bringing it to the position at which it forms an image of the utmost sharpness in the proper image plane. In accordance with the character of the *focusing mechanism* used for this act, the word may be qualified by the adjective coarse or fine.

focus, depth of *See* **depth of force**

focusable eyepiece *See* **eyepiece, focusable**

focusing magnifier *See* **magnifier, focusing**

focusing mechanism *See* **mechanism, focusing**

focusing mechanism (of the microscope) *See* **mechanism, focusing (of the microscope)**

focusing screen, clear A sheet of clear glass or plastic material used for focusing in photography and *photomicrography*. A figure on the screen (e.g. *cross lines*) serves to define the plane in which the *aerial image* observed with a *focusing magnifier* must be located.

Fourier, Jean B. J. (1768–1830) A French mathematician and physicist who developed wave theory.

fovea [physiol.] A pit in the centre of the macula of the *retina* which contains only *cones*. In the *light-adapted* state the fovea is the area which has an especially high sensitivity for *luminance* and *colour* differences; it is also that part of the retina with a high *visual acuity*.

Fraunhofer, Joseph von (1787–1826) A German physicist, discoverer of the lines in the solar spectrum.

Fraunhofer diffraction *See* **diffraction, Fraunhofer**

free working distance (of the objective), *See* **objective, free working distance of**

Fresnel, Augustin Jean (1788–1827) A French civil engineer who carried out fundamental research in optics, especially in interference, diffraction, and the wave theory of light.

Fresnel diffraction *See* **diffraction, Fresnel**

Fresnel lens A flat *lens* (usually moulded) from plastic) one surface of which does not possess a continuous curvature but is formed by a series of concentric ring-shaped steps. By suitably curving the different zones, the Fresnel lens may be made to act as an aspheric lens. Fresnel lenses are much lighter than conventional glass lenses and are mainly used in optical instruments as *field lenses*.

front focal plane *See* **focal plane, front**

front lens *See* **lens, front**

G

Gaussian space *See* **space, Gaussian**

geometrical optics *See* **optics, geometrical**

glare (1) Disturbance of the functioning of the *eye* (physiological glare) and/or the sensation of discomfort (psychological glare) caused by differences of *luminance* within the *field of view* or by very high luminance across the whole field of view. (2) Stray *light* which diminishes *contrast*.

glass, crown An optical glass characterized by low *refraction* and low *dispersion*.

glass, flint An optical glass characterized by high *refraction* and high *dispersion*.

glass, ground Glass, whose surface is roughened by mechanical or chemical means, which is used in microscopy to provide scattering or *diffusion* of the *light* passing through or falling on it. It may be used as a screen for the visualization of a *real image*.

glass, heat-absorbing A type of glass suitable for a *heat filter*.

glass, opal A type of glass within which *diffusion* takes place, giving it a white or milky appearance.

goniometer An instrument for measuring angles.

goniometer eyepiece *See* **eyepiece, goniometer**

graticule A pattern such as a scale or grid, together with its support, placed in an *object* or *image plane*. It is used for measurement, reference, alignment, location, counting, and stereological analysis.

graticule, eyepiece- A *graticule* which can be placed inside an *eyepiece*.

graticule eyepiece *See* **eyepiece, graticule**

grating, diffraction A set of structures repeating regularly in 1, 2, or 3 dimensions, which when illuminated produce, by *reflection* or *transmission*, maxima and minima of *intensity* as a consequence of *interference*. These maxima and minima vary in position according to *wavelength*. *Radiation* of any given wavelength may thus be selected from *complex radiation* allowing the grating to be used as a *monochromator*.

Greenough, Horatio An American zoologist who suggested that Zeiss make a binocular microscope with two separate viewing systems with their axes inclined and fitted with prisms to give an erect image so that the microscope could be used for dissection.

Greenough microscope *See* **microscope, stereo-**

ground glass *See* **glass, ground**

H

haemocytometer A *counting chamber* designed for enumerating blood cells.

half-peak-height-bandwidth *See* **bandwidth, half-peak-height**

half-wave plate *See* **plate, half-wave**

halo A phenomenon in *phase-contrast microscopy* by which a feature in the *image* is surrounded by a dark or light rim.

halogen lamp *See* **lamp, halogen**

hanging drop A drop of fluid containing particles, hanging from the underside of a *cover glass* and normally supported on a *hanging-drop slide*.

hanging-drop slide *See* **slide, hanging-drop**

head, camera A *tube head* in which a photomicrographic camera is incorporated or onto which a camera may be mounted. It is frequently combined with a *viewing tube* to allow simultaneous or alternate viewing and/or *photomicrography*. It may also contain means for determining and controlling the photographic exposure.

head, photometer A *tube head* equipped mechanically and optically to carry a photoelectric sensor. It may contain means for selecting and restricting the *photometric field*.

head, projection A *tube head* equipped with a *screen* and some means for forming a *real image* on it, so that the image can be observed without looking into an *eyepiece*.

Note: The screen may be replaced by a large-format camera, enabling the projection head to act as a camera head.

head, television A *tube head* equipped to carry a television camera.

head, tube *See* **tube head**

heat-absorbing glass *See* **glass, heat-absorbing**

heat filter *See* **filter, heat**

heating stage *See* **stage, heating**

hiatus The distance along the *optical axis* between the *principal planes* H and H′ of a *lens*. (See Fig. 4.) The hiatus is one of the *cardinal distances*.

high-eyepoint eyepiece *See* **eyepiece, high-eyepoint**

Homal eyepiece *See* **eyepiece, Homal**

homogeneous immersion *See* **immersion, homogeneous**

Hooke, Robert, (1635–1703) An English scientist and early user of the microscope. Published a book *Micrographia* (1665) in which the word ‘cell’ was first used to describe the cavities seen in a slice of cork.

hue [physiol.] That quality of *colour* subjectively described by the terms red, green, blue, etc. (*See* **wavelength, dominant**)

Huygenian eyepiece *See* **eyepiece, Huygenian**

hypermetropia [physiol.] Farsightedness. Defective vision in which the *far-point* is situated a finite distance behind the eye.

I

illuminance (at a point on a surface) [photom.] The incident *luminous flux* per unit area of surface. (See Table 1.)

Note: The use of the term illumination for this specific quantity conflicts with the more general usage of the word.

illuminated field *See* **field, illuminated**

illuminated field diaphragm *See* **diaphragm, illuminated field**

illuminating aperture diaphragm *See* **diaphragm, illuminating aperture**

illuminating system A *source* enclosed in a suitable housing together with a *collector*, a *condenser*, and supplementary *lenses* if required. Used to illuminate the *object* in a manner appropriate for the method of observation employed.

illumination (1) The application of *light* to an *object*, i.e. lighting it. (2) A term often loosely used for the *photometric quantity illuminance*.

illumination, annular Any arrangement of *illumination* in which the *rays* form a hollow cone.

illumination, axial *Illumination* with a *ray bundle* whose axis coincides with the *optical axis* of the *microscope*.

illumination, bright-field (1) A system of *illumination* for *bright-field microscopy*. (2) The kind of illumination supplied by this system.

* **illumination, 'critical'** *See* **illumination, source-focused**

illumination, darkground (1) A system of *illumination* for *darkground microscopy*, which may be qualified as *transmitted-light* or *reflected-light* darkground. (2) The kind of illumination supplied by this system.

illumination, epi- *Illumination* which falls on the *object field* from the same side as that from which the object field is observed. The axis of the illuminating *ray bundle* falling on the object may or may not coincide with the microscope axis.

illumination, incident *Illumination* which falls on the *object* from any direction.

Note: This term is sometimes used incorrectly to refer to epi-illumination.

illumination, Köhler A method of illuminating *objects* in which an *image* of the *source* is projected by a *collector* into the plane of the *aperture diaphragm* in the *front focal plane* of the *condenser*. This latter, in turn, projects an image of an *illuminated field diaphragm* at the opening of the collector into the *object plane*. In *epi-microscopy* (where the *objective* serves as its own condenser) an aperture diaphragm is imaged by a *relay lens* into the *back focal plane* of the objective and the illuminated field diaphragm is arranged to be in a plane *conjugate* with that of the collector.

* **illumination, Nelsonian** *See* **illumination, source-focused**

illumination, Rheinberg A method of *transmitted-light illumination* in which *darkground illumination* in one *colour* is combined with *bright-field illumination* in another colour. It is achieved by using a suitably designed *filter* in the *lower focal plane* of the *condenser*.

illumination, source-focused A method of *illumination* in which an *image* of the *source*, which may carry an *illuminated field diaphragm*, is projected by the *condenser* into the *object plane*. The source must be homogeneous in order to obtain even illumination. Synonymous with the obsolescent term 'critical illumination' which should no longer be used.

illumination, transmitted-light (or trans-illumination) *Illumination* which passes through the *object field*.

illumination, unilateral oblique *Illumination* using a *ray bundle* whose axis makes an angle with the *optical axis* of the *microscope*.

Note: Unilateral oblique illumination is often simply called oblique illumination.

illumination, vertical Axial *epi-illumination*.

illuminator, epi- Part of the *illuminating system* of the *reflected-light microscope* placed between the *objective* (which serves as its own *condenser*) and the *lamp* fitting. It is attached to or is inserted into the *body tube*, thus forming a section of this tube. A *reflector*, or a set of interchangeable reflectors, is included in the illuminator.

illuminator, fibre optic A *microscope lamp* in which the *light* is delivered by a *fibre optic*. This may be flexible, it may branch, may terminate in *lenses*, or be arranged as a *ring light*.

image A structural representation of those properties of the *object* which cause modulation of *light*. All parameters which describe the spatial and the temporal state of light can be modulated. Because of these modulations the information about the object is carried by the light in an encoded form. In *geometrical* *optics* the image is the appearance in that plane in the *image space* of a *lens* (or other imaging system) at which the individual image points are *conjugate* with the corresponding points in *object space*. In *microscopy* with the *compound microscope* a *primary image* and a secondary image are formed, the latter being produced on the *retina* of the observer's *eye* or on photographic material or other surface.

image, aerial A *real image* existing in a plane in space and not normally visible to the naked eye.

image, bright-field An *image* formed by *bright-field illumination*.

image, darkground An *image* formed by *darkground illumination*.

image, primary The (usually) magnified *real image* of the *object* formed by the *objective*. This must not be confused with the 'primary interference image' as described by Abbe.

image, real An *image* which can be received on a surface, e.g. a screen.

image, retinal Any *real image* produced on the *retina* of the *eye* by the optical system of the eye alone or by that in co-operation with an external optical system.

image, virtual An *image* which cannot be received on a surface but which may be converted into a *real image* by the optical system of the *eye* or other *converging lens* system.

image, analysis [quant.] The separation and evaluation of an *image* into its compositional and textural elements. It includes the description of the image with the help of optical, geometrical, and stereological parameters and, optionally, the elucidation of the general principles underlying individual compositional and textural phenomena.

image converter A device by which an *image* formed (usually by invisible *radiation*) on a photoemissive or photoconductive surface produces a corresponding image on a luminescent surface. A television camera equipped with a special tube or other sensitive device may be used in this way. Image converters are used to provide visible images by amplification from an original image of low intensity, or from images formed by *radiation* to which the *eye* is not sensitive.

image distance *See* **distance, image**

image distance correction of objective *See* **objective, image distance correction of**

image field *See* **field, image**

image field, curvature of The formation of a curved *image field* from a plane *object field*. It is particularly obvious with *objectives* of high *magnifying power* and large *numerical aperture*, which have a restricted *depth of field*. It may largely be eliminated by additional *correction*. (*See* **eyepiece, flat-field; objective, flat-field**).

image intensifier An *image converter* used to provide a visible *image* in conditions of low *light* levels.

image plane *See* **plane, image**

image-side aperture (of a microscope objective) *See* **aperture, image-side**

image space *See* **space, image**

imaging depth *See* **depth of focus**

imaging mode, confocal A technique in which a *point source* is imaged into a pinhole is placed in front of the photodetector of a *scanning optical microscope*, resulting in a strong optical sectioning property which can be used to give three-dimensional *images*. *Resolving power* is improved and stray light rejected.

imaging scale *See* **magnification, lateral**

immersion The use of an *immersion liquid*.

immersion, homogeneous *Immersion* in which the *immersion liquid* and the adjacent optical components have the same *refractive index* and *dispersive power* (or *Abbe number*) so that no *refraction* occurs between the liquid and the optical components.

immersion, homogeneous, of objective (or condenser) *Homogeneous immersion* between the preparation and the *objective* and/or *condenser*.

Note: In modern microscope design, however, refractive index differences between objective front lens, the immersion liquid, and the cover glass are deliberately introduced in order to assist in the correction of the system so that the immersion is not homogeneous.

immersion lens *See* **lens, immersion**

immersion liquid A liquid (commonly *immersion oil*, water, or glycerol) specified as suitable for use in the space between the front of an *immersion lens* and the *object*. Because the immersion liquid is considered in the computing of *corrections* to be part of the *lens*, its *refractive index* and *dispersive power* (or *Abbe number*) are critical. The optical data for such liquids are standardized.

immersion oil Currently used to refer to a synthetic *immersion liquid* now governed by the ISO Standard 8036/1; formerly applied to naturally occurring immersion liquids such as cedar wood oil.

incandescence The *emission* of visible radiation by thermal excitation.

incidence The falling of *light* onto a surface from any direction.

incidence, angle of The angle between the normal to a surface and a *ray* or the axis of a *beam* falling on that surface.

incident beam *See* **beam, incident**

incident illumination *See* **illumination, incident**

incident ray *See* **ray, incident**

incoherent An adjective expressing the property of *light* not satisfying the *coherence condition* and thus incapable of *interference*.

index of refraction *See* **refractive index**

indicatrix [pol.] A spatial pattern or figure in the shape of an ellipsoid of revolution or triaxial ellipsoid. It represents the *refractive indices* as distances between the centre of the ellipsoid and points on its surface and the *vibration directions* of *plane-polarized light* related to the refractive indices by the directions of the said distances.

Note: Ellipsoids of revolution are related to optically uniaxial media, triaxial ellipsoids are related to optically bixial media.

infinity-corrected objective *See* **objective, infinity-corrected**

infra-red microscopy *See* **microscopy, infra-red**

infra-red radiation *See* **radiation, infra-red**

integrating eyepiece *See* **eyepiece, integrating**

integrating stage *See* **stage, integrating**

intensity A general term for the strength of a *radiation*. It is proportional to the square of the *amplitude* of the *electromagnetic wave*.

Note: For measurement, this term should be replaced by the most suitable photometric or radiometric quantity.

intensity, luminous (of a source) [photom.] The *luminous flux* emitted about a given direction per unit of *solid angle*. (See Table 1.)

intensity, radiant (of a source) [photom.] The *radiant flux* emitted about a given direction per unit *solid angle*. (See Table 1.)

intensity contrast *See* **contrast, photometric**

interference The mutual interaction between two or more *coherent wavetrains*. The phenomenon is used to convert *optical pathlength differences* in the *object* into *intensity* variations in the *image* so providing *contrast*.

interference, differential *Double-beam interference* in which two *wavetrains* which fall on the *object plane* or *image plane* are separated laterally by a distance similar to the *minimum resolvable distance*. This kind of interference is characterized by *contrast* which gives an impression of *unilateral oblique illumination*. In *transmitted light*, variations in *optical pathlength* which are not due to differences in physical thickness also appear as *relief* in the *image*.

interference, double-beam *Interference* in which *wavetrains* from two *light beams* are involved. In practice the two beams are derived from a single *source*.

interference, double-focus *Double-beam interference* in which the two *light beams* have different levels of *focus*. One beam is focused in the *object plane*, the other above or below that plane.

interference, multiple-beam *Interference* in which *wavetrains* from more than two *light beams* are involved.

interference, polarizing *Double-beam interference* in which the two *light beams* arise from one *plane-polarized beam* as the result of *double refraction*. They are also *plane-polarized* in mutually perpendicular *vibration directions* and are recombined by the *analyser*.

interference, shearing *Double-beam interference* in which the two *light beams* which fall upon the *object plane* or *image plane* are separated laterally from one another. This separation limits the size of features which can be studied.

interference colour *See* **colour, interference**

interference contrast (1) A term used in *interference microscopy* to indicate that the *contrast* in the *image* is mainly caused by *interference*. (2) The phenomenon of *interference* as used for enhancing the contrast between features having different *optical pathlengths*.

interference filter *See* **filter, interference**

interference image, primary A term introduced by Abbe for the *primary diffraction pattern*.

interference microscope *See* **microscope, interference**

interference microscopy *See* **microscopy, interference**

interferometry The use of *interference* phenomena to make measurements, mainly of *optical pathlength differences*. *Refractive indices* and thickness can also be derived.

interferometry, microscope The application of *interferometry* to the investigation of microscopic objects which are usually observed at the same time.

intermediate lens *See* **lens, intermediate**

intermediate tube *See* **tube, intermediate**

internal-diaphragm eyepiece *See* **eyepiece, internal-diaphragm**

internal transmission density *See* **absorbance**

interpupillary distance *See* **distance, interpupillary**

intersection distance *See* **distance, intersection**

inversion [physiol.] The alternation between two possible interpretations or a correct and an incorrect interpretation of an *image*, e.g. the impression of *relief*. It occurs when the brain misinterprets the similarity of the image to another image previously experienced. Inversion arises chiefly because of missing geometrical perspective and/or misinterpretation of the distribution of bright and dark areas as if they were light and shadow. In *microscopy* inversion may occur when *azimuthal* contrasting methods such as unilateral *oblique illumination* or *differential interference contrast* are used.

inverted microscope *See* **microscope, inverted**

iris [physiol.] A circular sheet of tissue, located in front of the *lens* of the human *eye*. The iris controls the diameter of the *pupil* of the eye, which is formed by the central circular opening of the iris.

iris diaphragm *See* **diaphragm, iris**

irradiance [photom.] The *radiant flux incident* upon a unit area of the surface (see Table 1).

irradiation The application of *radiation* to an *object*.

ISO The abbreviation for the International Organization for Standardization, the international authority for technical standards.

isochromatic curves [pol.] *Interference*

bands seen in the *back focal plane* of the *objective* or in any succeeding plane *conjugate* with it during *conoscopic observation*. The bands show *interference colours* of continuous orders, indicate spatial directions of equal *optical path-length difference* in the *object* and intersect the *isogyres*.

Note: With monochromatic light the coloured bands are seen as alternating dark and bright bands.

isogyres [pol.] Dark straight or curved brushes seen during *conoscopic observation*.

Note: The brushes are intersected by the isochromatic curves and, together with these, form the conoscopic interference figure. Isogyres are related to extinction curves.

isotropy (1) The uniform spatial distribution of properties of a material. (2) In *polarized-light microscopy* usually taken to refer to the uniformity of optical properties with respect to the *vibration plane* of *polarized light*.

K

Kellner eyepiece *See* **eyepiece, Kellner**

Köhler, August (1866–1948) A German biologist who joined Carl Zeiss in 1900 and developed the method of microscopical illumination which bears his name. He also worked in ultraviolet microscopy with von Rohr and designed many microscopical accessories, including (together with Hans Boegehold) the Homal eyepiece.

Köhler illumination A method of illuminating *objects* in which an *image* of the *source* is projected by a *collector* into the plane of the *aperture diaphragm* in the *front focal plane* of the *condenser*. This latter, in turn, projects an image of an *illuminated field diaphragm* at the opening of the collector into the *object plane*. In *epi-microscopy* (where the *objective* also serves as its own condenser) an aperture diaphragm is imaged by a *relay lens* into the *back focal plane* of the objective and the illuminated field diaphragm is arranged to be in a plane *conjugate* with that of the collector.

L

lamp A source of artificial *light*. A *bulb*.

lamp, discharge A *lamp* producing *complex radiation* by an electrical discharge through a rare gas or the vapour of a metal. The emitted *spectrum* is composed of lines or bands, usually superimposed on a continuous background.

lamp, filament A *lamp* from which *complex radiation* is emitted from a filament, usually of tungsten, heated by the passage of an electric current. The emitted *spectrum* is continuous and approximates to that of a *black-body radiator*.

lamp, halogen A *filament lamp* whose *quartz envelope* contains halogen vapour. Loss of tungsten from the filament and its deposition on the envelope is reduced by a cyclical process involving the halogen. This permits a high filament temperature and consequent higher *luminance*, higher *colour temperature*, and longer operating life than a conventional filament lamp of the same input power.

lamp, mercury arc A *discharge lamp* containing mercury vapour, often at a high pressure when the lamp is operating. At low pressure the lamp emits a characteristic *line spectrum* but when it heats up there is a strong *continuous spectrum* forming the background. This type of lamp is frequently used in *fluorescence microscopy* and also, with a suitable *filter*, as a source of *monochromatic light* or *ultraviolet radiation*.

lamp, microscope The *lamp* together with its housing and fittings. It may be a separate unit or be incorporated into the *microscope* itself.

lamp, solid-source A *filament lamp* in which the tungsten filament is closely wound, often on a flat mandrel, so that the coils are almost in contact to provide a *primary source* of relatively large area and uniform *luminance* across its area.

lamp, xenon arc A *discharge lamp* containing xenon, often at a high pressure when the *lamp* is operating. The lamp emits *light* of high *luminance*, high *colour temperature* and with an almost continuous *spectrum* distributed from the *ultraviolet* to the *infra-red*.

Land, Edwin H. (1909–] An American inventor and physicist. In 1932 he used aligned submicroscopic crystals of iodoquinine embedded in a sheet of plastic as a polar. This was marketed as Polaroid. Later he invented a one-step process for developing and printing photographs and founded the Polaroid corporation.

laser An acronym for Light Amplification by Stimulated Emission of Radiation. A *source* which emits *coherent radiation* of high *spectral concentration* of *radiance* and an extremely small *solid angle*.

lateral chromatic aberration *See* **aberration, chromatic, lateral**

lateral magnification *See* **magnification, lateral**

law, Beer–Lambert *See* **Beer–Lambert law**

law, Snell's *See* **Snell's law**

Leeuwenhoek, Antony van (1632–1723) A Dutch linen-draper and one of the first microscopists who constructed many single lens microscopes of extraordinary resolving power.

lemniscate [pol.] A figure-of-eight shaped curve, descibing the appearance of one type of *conoscopic interference figure*.

lens A piece of transparent material with one or more curved surfaces, which is used to alter systematically the direction of *rays* of *light*.

Note: The term lens may also be used for a system of lenses which, in principle, acts as a single lens.

lens, aspherical A *lens* made with an *aspherical* surface; in *microscopy* sometimes found in *illuminating systems*.

lens, Bertrand (Amici–Bertrand) An *intermediate lens* which transfers an image of the *back focal plane* of the *objective* into the *primary image plane*. Used for *conoscopic* observation in *polarized-light microscopy* and for adjustment of the microscope *illuminating system*, especially with *phase-contrast microscopy*.

lens, collector *See* **collector**

lens, converging A *lens* which is thicker at its centre than at its edge and which can produce a *real image*. It increases the convergence or decreases the divergence of a *light-beam*. The same term is often applied to a lens system which acts in the same way as a single converging lens.

lens, diverging A *lens* which is thinner at its centre than at its edge and which, when used alone, can only produce a *virtual image*. It increases the divergence or diminishes the convergence of a *light beam*. The same term is often applied to a lens system which acts in the same way as a single diverging lens.

lens, eye- *See* **eyelens**

lens, field *See* **field lens**

lens, front Commonly used with reference to an *objective*, where it is that part of the objective which is nearest to the *object*.

lens, immersion An *objective* or *condenser* designed to work with an *immersion liquid*.

lens, intermediate A special name for a *relay lens* located between the *objective* and the *primary image*. It serves to control the position, quality, and/or *lateral magnification* of the primary image and/or to restore the conditions for correct optical imaging, if the actual *optical fitting dimensions* are different from the standard ones.

lens, meniscus A *lens* which is concave on one surface and convex on the other.

lens, negative *See* **lens, diverging**

lens, positive *See* **lens, converging**

lens, projection A *lens* designed to form a real *image* at a finite distance; it is used for projection, drawing or photomicrographic purposes. The projection distance may be varied by a focusing device.

lens, relay (1) A *lens* used to link together the *light paths* in optical devices. (2) A lens used to adapt the light path either on the illumination or the image side to given mechanical and/or optical requirements. (*See* **lens, intermediate; tube-lens**)

lens, thick A *lens* or system of lenses in which the distance between the outer faces is large compared with the *focal length*, and in which the *principal planes* are separated by a *hiatus*.

lens, thin A *lens* in which the distance between the outer faces is small compared with the *focal length*, and in which the *principal planes* coincide.

lens, top Commonly used to refer to the top *lens* of the *condenser* which is that part of the condenser nearest to the *object*.

lens, tube- *See* **tube-lens**

lens, tubelength correction An *intermediate lens* used to correct optically any deviation of *mechanical tubelength* from its nominal value.

lens mount *See* **mount, lens**

lens of the eye [physiol.] A firm, gelatinous biconvex body consisting of many layers of fibres which are derived from cells and are of slightly differing *refractive indices*. It is located behind the *pupil of the eye*. Its *focal length* is variable (between limits) by muscular control of its thickness and shape: thus it (together with the *refractive power* of the *cornea*) can form clear images on the *retina* of objects at different distances from the eye. (*See* **accommodation**)

levelling (of a polished section) The mounting of a *polished section* so that its surface is normal to the *optical axis* of the *microscope*.

levelling stage *See* **stage, levelling**

Lieberkühn, Johannes Nathaniel (1711–56] A German anatomist who used a simple microscope fitted with a reflector to study injected animal specimens by epi-illumination.

Lieberkühn (Lieberkühn reflector) A concave mirror fitted around the *objective* and serving to reflect light from beneath the *stage* down onto the upper surface of an *object*.

light (visible radiation) An electromagnetic *radiation* capable of directly causing a visual sensation. The limits of visible radiation are not well defined but are usually taken as being from 380–400 nm to 760–80 nm. In microscopy, *infra-red* and *ultraviolet* radiations are sometimes included with *light*, even though they do not directly excite a visual sensation.

light, circularly polarized [pol.] A special case of *elliptically polarized light* in which there is a phase difference of 90 degrees (or an odd integer multiple of this figure) between two superimposed *wavetrains* of equal *amplitude* having their *vibration directions* at right angles to one another. (*See also* **plate, quarter-wave**)

light, diffracted *See* **diffracted light**

light, direct *See* **direct light**

light, elliptically polarized [pol.] *Light* which results from the superimposition of at least two *plane-polarized wavetrains* having their *vibration directions* at right angles to each other. They are not in phase but have a fixed phase relationship and equal or different *amplitudes*. The resulting *light vector* propagates in the form of a helix on the surface of an elliptical cylinder.

light, natural *Light* containing a multiplicity of *wavetrains* of different *wavelengths*, *amplitudes*, and *polarization states*, travelling in the same direction.

light, plane–polarized *Light* in which there is only one *vibration direction*.

light, polarized *Light* in which the vibrations are partially or completely suppressed in certain directions.

light, ring A circular *source* of *light* or *fibre optic illuminator* which surrounds a *lens* or *optical axis*. Used to provide multi-directional *illumination*.

light, stray *Light* which arises from scatter or *reflection* by the *object*, in *lenses*, or by obstacles in the *light path*. It does not contribute to *image* formation but reduces the *contrast* in the image.

light, white *Light* containing any mixture of *monochromatic radiation* in such proportions that it will be perceived as being without *chromatic colour*.

light adaptation [physiol.] The process of adjustment of the *visual organ* from lower to higher *luminance*.

light adaptation, state of [physiol.] The final state of *adaptation* in *daylight vision* (*photopic vision*).

light beam *See* **beam, light**

light exposure *See* **exposure, light**

light filter *See* **filter**

light microscopy *See* **microscopy, light**

light path *See* **path, light**

light source *See* **source**

light vector *See* **vector, light**

limb That part of a *microscope stand* which carries the principal parts of the instrument (see Fig. 1).

line, fixation [physiol.] A straight line joining the centre of the *entrance pupil* of the observer's *eye* to the *fixation point* of the *object* viewed. The angle between the fixation lines of the two eyes is the convergence angle of the eyes. If the object point is located at infinity, the fixation lines are parallel; in this case their separation is the *interpupillary distance*.

line spectrum *See* **spectrum, line**

linear magnification *See* **magnification, line**

Linnik, Vladimir Pavlovich (1889–) A Russian physicist involved in optical research. Developed the principle of a microinterferometer which bears his name.

Lister, Joseph Jackson (1786–1869) A London wine trader who was interested in optics and microscopy. He developed the principle of aplanatic focal points which was fundamental to the improvement of the compound microscope.

locating flange of eyepiece *See* **eyepiece, locating flange of**

locating flange of objective *See* **objective, locating flange of**

locating surface (or flange) The surface (or flange) at which two interchangeable components fit together. These surfaces (or flanges) are perpendicular to the *optical axis* and are responsible for setting the correct axial location and centration of the optical and mechanical elements. They may coincide with *reference planes* for *optical fitting dimensions*. (*See* Figs. 1 and 2.) Some or all of the following locating surfaces may be found:

Locating flange of eyepiece	Lower locating surface of accessory
Eyepiece-locating surface of viewing tube	Upper locating surface of tube head
Lower locating surface of viewing tube	Lower locating surface of tube head

Upper locating surface of intermediate tube
Lower locating surface of intermediate tube
Upper locating surface of body tube
Objective-locating surface of nosepiece
Locating flange of objective

locating surface for eyepiece *See* **eyepiece-locating surface (of the viewing tube)**

locating surface for objective (of the nosepiece) *See* **objective-locating surface (of the nosepiece)**

locus, spectrum *See* **spectrum locus**

long-wave-pass filter *See* **filter, long-wave-pass**

longitudinal magnification *See* **magnification, axial**

loupe *See* **magnifier**

lumen [photom.] The SI unit of *luminous flux*; the amount of *light* emitted in a unit *solid angle* by a point source of output one *candela*. Alternatively, it is the amount of light which falls on a unit area when the surface is at a unit distance from a source of intensity one candela.

luminance [photom.] (1) The quality or state of being luminous (i.e. emitting *light*). (2) *Luminous flux* leaving, arriving at or passing through a unit area of surface per unit *solid angle*. (*See* Table 1.)

luminous factor [photom.] The *luminance* with which a reflecting body appears to an observer, expressed as a proportion of the luminance of a perfectly reflecting *white body*.

Note: The factor is determined by the spectral luminous efficiency.

luminescence *See* **photoluminescence**

luminosity The subjective assessment of *brightness*.

luminosity curve [physiol.] The graph showing the *spectral distribution* of the *spectral luminous efficiency of the eye* (the $V(\lambda)$ curve).

luminous efficiency *See* **efficiency, luminous**

luminous exitance *See* **exitance, luminous**

luminous flux *See* **flux, luminous**

luminous intensity *See* **intensity, luminous**

luminous transmittance [photom.] The *luminance* with which a transparent body appears to an observer, expressed as a proportion of the luminance of a perfectly transparent *white body*.

Note: Luminous transmittance is governed by spectral luminous efficiency.

lux [photom.] The derived SI unit of *illuminance*. One lux = one lumen per square metre. (*See* Table 1.)

M

M A symbol sometimes used for *magnifying power*; may be qualified M_{ang} or M_{lat} for angular or lateral magnifying power respectively.

* **macrophotography** *See* **photomacrography**

magnification (1) The degree by which dimensions in an *image* are, or appear to be, enlarged with respect to the corresponding dimensions in the *object*. (2) The act or process of enlarging.

Note: The use of the word magnification alone should be avoided unless the distinction between angular, lateral, and other forms of magnification is clear from the context.

magnification, angular The ratio between the tangent of the *viewing angle* of an *object* when observed through a magnifying system and the tangent of the viewing angle of the same object when observed by the naked *eye* at the *reference viewing distance*. This ratio should be expressed as for example, ×10.

Note: For a magnifier or eyepiece, angular magnification may be calculated from the expression

$$M_{ang} = \frac{\tan \sigma'}{\tan \sigma} = \frac{\text{reference viewing distance (250 mm)}}{\text{focal length of magnifier (mm)}}$$

(See Fig. 7b).

For the purpose of this calculation the object is taken to extend from the optical axis on one side only.

magnification, areal The ratio between the size of an area in the *image* and the size of the corresponding (*conjugate*) area in the *object*.

magnification, axial The ratio between a given axial distance in *object space* and the corresponding (*conjugate*) distance in *image space*.

magnification, empty Any *magnification* greater than the *useful magnification range*; exceeding this range gives no further information about the *object*, but sharpness and *contrast* appear to decrease.

magnification, lateral *Linear magnification* normal (perpendicular) to the *optical axis*. This ratio should be expressed in a proportional form, e.g. 10:1. (See Fig. 7a).

magnification, linear The ratio between a given distance in a *real image* and the corresponding (*conjugate*) distance in the *object*.

magnification, photomicrographic The *lateral magnification* of a *photomicrograph*. Its value is the product of the *magnifying powers* of the *objective* and *eyepiece* (when the eyepiece is used as a *projection lens*), the *tube factor*, the camera factor, and the degree of photographic enlargement. It should be expressed in a proportional form, e.g. 10:1, or a *scale bar* drawn on the *micrograph*.

magnification, primary The *magnifying power* of an *objective*.

magnification, total The *magnifying power* of the entire imaging system of the *microscope*.

magnification, useful range of That range of *total magnifications* within which details in the *object* are clearly seen in the *image*. The value of this range is usually taken to lie between 500 and 1000 times the *numerical aperture* of the *objective*. When the effective total magnification is less than the lower limit, the *resolving power* of the objective cannot be fully utilized; when the magnifying power exceeds the upper limit *empty magnification* occurs.

magnification changer An *intermediate lens* for changing the *lateral magnification* of the *primary image*. Its effect is expressed as a *tube factor* which may be varied step by step or continuously. In the case of an *infinity-corrected objective* the same effect may be achieved by exchanging the *tube-lens* for another of different *focal length*.

magnification number of objective *See* **magnifying power of objective**

magnifier A *converging lens* used between the *object* and the *eye* to increase the *viewing angle* and hence to provide a magnified *image* on the *retina* of the eye.

magnifier, focusing An adjustable *magnifier* used to help in the precise focusing of an *image* in photography and photomicrography.

magnifying power The ability of an optical system to produce a magnified *image* under specified working conditions (for example the *optical fitting dimensions*). The magnifying power is expressed as the *lateral* or *angular magnification* of the image under consideration.

Note: Magnifying power is often referred to as magnification.

magnifying power of eyepiece The *angular magnification* involved in the formation of a *virtual image* from the *primary image* by an *eyepiece*. Its value is indicated by, for example, × 10.

magnifying power of microscope The *angular magnification* involved in the formation of a *virtual image* by a *microscope*. This *magnification* is the product of the *magnifying powers* of the *objective*, the *eyepiece*, and any *tube factor(s)*.

magnifying power of objective The *lateral magnification* of the *primary (real) image* formed under specified operating conditions by an *objective* alone (when this *lens* is corrected for a finite *image distance*) or by the objective and the *tube-lens* of appropriate focal length (when an *infinity-corrected objective* is involved). Its value is indicated by the magnification number, e.g. 10. When quoted for infinity-corrected objectives without their tube-lens, the magnifying power is expressed in the form × 10.

marginal contrast *See* **contrast, marginal**

marginal ray *See* **ray, marginal**

marker, object An accessory which may be fitted to the *nosepiece* which, when moved in place of the *objective*, will mark an area of interest on an *object* or *preparation*.

marking (of optical components) The inscribing of data in the form of characters or colour bands on to optical components in order to indicate their optical properties, values of certain properties, and the origin of a component.

marking, colour, of objectives *See* **colour marking of objectives**

mask An opaque *diaphragm* in the plane of an *image*, serving to define the size and shape of the *image field* viewed or recorded.

maximum spectral luminous efficacy (of the eye) *See* **efficacy (of the eye), maximum spectral luminous**

measuring microscope *See* **microscope, measuring**

mechanical stage *See* **stage, mechanical**

mechanical tubelength *See* **tubelength, mechanical**

mechanism, focusing The mechanism used to change the distance between the *object* and the optical system forming an *image*, with the purpose of obtaining maximum sharpness in that image.

mechanism, focusing (of the microscope) The *focusing mechanism* (often mediated by a rack and pinion) which converts the rotary motion of a knob into linear motion along the *optical axis* of either the *objective* (with or without the *tube*) or the *stage*.

medium (1) Any substance. (2) A transparent substance through which electromagnetic *radiation* passes. (3) A short form of the term *mounting medium*.

mercury arc lamp *See* **lamp, mercury arc**

mesopic vision *See* **vision, twilight**

metallography The study of the microscopical structure of metals and alloys.

meter, exposure *See* **exposure meter**

microdensitometer A *photometer* equipped with a mechanism for scanning a microscopical *object* or the emulsion of photographic negatives and arranged for the measurement of *optical density* over a small but finite area.

micrograph A photographic record of an *image* formed by a *microscope*.

microhardness tester A device for measuring the hardness of *microscopic* areas of a *specimen*, for example individual grains in a metal surface. It consists of a steel sphere (in the Brinell system) or a pyramidal diamond point which can be impressed into the test area by a known force. The area of the indentation thus formed is measured and used as a basis for calculating the hardness.

microinterferometer An instrument specially designed for interferometry on small areas or surfaces. It does not allow visual inspection as with a normal *microscope*.

micromanipulator An instrument which allows fine manipulation of components of a *preparation* whilst they are observed with a *microscope*. This is achieved by means of mechanical reduction of the movements of the hand.

micrometer A device for measuring small lengths.

micrometer, filar *See* **eyepiece, micrometer-screw**

micrometer, stage A special *graticule* in the form of a scale carried at natural size on a *microscope slide*. It is used as an absolute standard of length for calibrating microscope measuring systems.

micrometer eyepiece *See* **eyepiece, micrometer**

micrometer-screw eyepiece *See* **eyepiece, micrometer-screw**

micrometre An SI unit of length equal to 1×10^{-6} metre (i.e. 0.001 mm). Symbol μm.

* **micron** An obsolete term for *micrometre*.

microphotography Photography, especially of documents, arranged to produce small images which cannot be studied without *magnification*. Not to be confused with *photomicrography*.

microphotometer A device for *photometry* of small specimens which does not, however, allow visual inspection of the *object* as with a normal *microscope*.

microphotometry The application of *photometry* to the investigation of small *specimens* which are not observed visually.

microprojector A *microscope* designed or adapted to project an enlarged *image* onto a *screen* for demonstration or drawing.

microscope An instrument designed to extend man's visual capability, i.e. to make visible minute detail that is not seen with the naked *eye*.

Note: The word is qualified by prefixes (electron, X-ray, acoustic, field-ion, etc.) unless it is clear from the context that the imaging involved is by means of light.

microscope, binocular A *compound microscope* in which a separate *image* is presented to each of the observer's *eyes* simultaneously. There are two types of binocular microscope: those in which, by the use of a special *viewing tube* and *beam splitter*, both eyes are presented with identical images, and *stereomicroscopes*.

microscope, comparison An arrangement of two *microscope* systems linked by a system of *prisms* to present their *images* into a *comparison eyepiece*. The *image field* is split so that the image from each microscope is seen in the corresponding half of the field. Used, especially in forensic science, to compare the fine details of two similar *specimens*.

microscope, compound A *microscope* which provides *magnification* in two stages by means of an *objective* and an *eyepiece*.

microscope, epi- See **microscope, reflected-light**

microscope, exit pupil of An area lying in a plane several millimetres after the *eyepiece* on the observer's side where an *image* of the objective's *exit pupil* is formed by the eyepiece together with any *tube-lenses* present. (See Figs. 8 and 9).

Note: The exit pupil of the microscope is important because its position and size dictate the position of the pupil of the observer's eye and the nature of other succeeding optical systems, such as cameras. Also termed the eyepoint or Ramsden disc.

microscope, field *See* **microscope, portable**

microscope, fluorescence A *microscope* specially designed or additionally equipped for *fluorescence microscopy*.

microscope, flying spot An early form of *scanning optical microscope* in which the intense spot of *light* forming the raster of a small cathode ray tube was imaged into the *object* plane of a *microscope* through an *eyepiece* and *objective*. A photodetector following the *condenser* received light transmitted by the *specimen* and modulated the *brightness* of the display cathode ray tube which was synchronized with the scan. It was used for counting small particles.

microscope, Greenough The original design of low-power *stereomicroscope*. It consists of two separate *compound microscope* systems mounted with their axes converging at an angle of about 15 degrees, so that they observe a common *field*. *Prisms* are fitted to erect the *image* and usually incline the *viewing tubes*. It provides an excellent stereoscopic image.

microscope, interference A *microscope* specially designed or additionally equipped for *interference microscopy* and/or *microscope interferometry*. Many types exist, often designated by the name of their originator, e.g. Jamin–Lebedeff, Linnik, Mach–Zehnder, Smith–Baker.

microscope, inverted A microscope, in which the *object* is observed from beneath the *stage*

microscope, light A *microscope* which uses *light* as the illuminating agent.

microscope, measuring A *microscope* specially designed with a calibrated lateral movement used to determine the distance between two points on an *object*.

microscope, monocular A *microscope* which presents the *image* to only one *eye*.

microscope, ore A combined *polarized-light* and *reflected-light microscope* specially used in the study and identification of ores, minerals, and other opaque *optically anisotropic* materials.

microscope, polarized-light A *microscope* specially designed or additionally equipped for *polarized-light microscopy*. It requires a *polarizer, analyser*, strain-free *lenses* between the *polars*, a rotable *stage* equipped with a scale to measure rotation angles, a mechanism for centration of the *objectives*, and a *focusable eyepiece* with centred and orientated *cross lines*. There is also a *Bertrand lens* and a tube slot for the insertion of *retardation plates* and *compensators*. A polarized-light microscope for reflected light is generally known as an *ore microscope*.

microscope, portable A compact *microscope* specially designed to be easily transportable and usable in the field.

microscope, projection *See* **microprojector**

microscope, reflected-light A *microscope* designed for the study of (mainly specularly) reflecting *objects*. It is equipped with an *epi-illuminator*.

Note: The preparations studied are mostly in the form of polished sections.

microscope, reflecting A *microscope* in which the *objective* and often the *condenser* systems are *catoptric*. Such systems are completely free from *chromatic aberration* and suffer less from *spherical aberration* than *dioptric* systems working under similar conditions. Practical difficulties have limited the production and use of reflecting microscopes.

microscope, scanning optical A *microscope* specially designed to scan the *object plane* or *image plane* in a *raster pattern*. *Light* signals at discrete and uniform intervals are received from the *object* by a photoelectric sensor and displayed on a screen or stored for further processing. The *image* is thus built up serially. There are two methods of scanning: one is based on movement of the illuminating *beam* with the object remaining stationary, the other on the movement of the object, the beam remaining stationary. The instrument may be operated in the *confocal imaging mode*.

microscope, simple A *microscope* consisting of only one *lens*, the *objective*.

microscope, stereo- A *binocular microscope* in which the *object* is observed by each *eye* from a slightly different angle. Disparate image points will be imaged on corresponding points of the *retina* and thus cause stereoscopic perception. The *Greenough* type of *microscope* has two completely separate optical systems inclined at a particular convergence angle with respect to each other with an inverting *prism* to give an erect image. More recent systems use a common main *objective* whereby the *convergence angle* of both paths of rays is achieved by dividing the *pupil* in the *back focal plane* of the objective.

microscope, television A *microscope* adapted so that its *image* is presented to the observer by a television system. Especially useful for demonstration. It may form part of *image analysis* equipment.

microscope, transmitted-light A *microscope* designed for the study of transparent *objects* by *transmitted light*.

Note: The preparation is normally mounted on a slide and may or may not be covered with a cover glass, according to the requirements of the technique.

microscope, travelling *See* **microscope, measuring**

microscope, video-enhanced contrast A special form of *television microscope* in which the *image* is electronically processed in order to enhance *contrast*. Especially valuable for the study of very small features of low contrast, such as occur in living cells.

microscope base *See* **base, microscope**

microscope fluorimetry *See* **fluorimetry, microscope**

microscope interferometry *See* **interferometry, microscope**

microscope lamp *See* **lamp, microscope**

microscope limb *See* **limb**

microscope mirror In the older types of *microscope*, a plane or concave mirror mounted in gimbals beneath the *stage*, serving to reflect *light* from an external source into the *optical axis* of the microscope. The concave surface is used with low-power *objectives* operated without a *condenser*.

microscope photometer A *microscope* equipped for *photometry*. It is called a microscope spectrophotometer if it is combined with a *monochromator*.

microscope photometry *See* **photometry, microscope**

microscope slide *See* **slide**

microscope stage *See* **stage**

microscope stand *See* **stand**

microscope tube *See* **tube**

microscopic Invisible or indistinguishable except with the aid of a *microscope*; very small.

microscopical Pertaining to a *microscope* or to *microscopy*.

microscopy The use of or investigation with a *microscope*.

microscopy, bright-field *Microscopy* in which the *direct light* passes through the *objective aperture* and illuminates the background against which the *image* is seen.

microscopy, darkground (darkfield)*Microscopy* in which the *direct light* is prevented from passing through the *objective aperture*. The *image* is formed from light scattered by features in the *object*, the detail thus appearing bright against a dark background.

microscopy, dispersion staining The *microscopy* of transparent objects which are in a *mounting medium* the *refractive index* of which matches that of the *object* for a certain *wavelength*, but which has a distinctly higher *dispersive power* than the object. Under these conditions both the object and the mounting medium appear coloured near their interfaces. The *colour* with which the object appears is distinctly different from that with which the mountant appears. The colours and their differences depend on the wavelength at which the refractive indices of the object and the medium match and the kind of microscopy used; dispersion staining may be used in *bright-field microscopy*, the colour being concentrated in the *Becke line*, in *darkground microscopy* or in *phase-contrast microscopy*.

microscopy, epi- *See* **microscopy, reflected-light**

microscopy, fluorescence *Microscopy* in which the *image* is formed by *fluorescence* emitted from the *object* itself. The object may be regarded as self-luminous and the *light* emitted is *incoherent*.

microscopy, infra-red *Microscopy* in which the *image* is formed by *infra-red radiation* and is displayed by means of an *image converter* or photographic film. Microscopy using near-infra-red radiation may be performed with a conventional *microscope*; microscopy in the far infra-red requires special equipment.

microscopy, interference *Microscopy* in which *physiological* and *photometric contrast* in the *image* is produced or influenced by the action of optical components which regulate *interference* in special ways.

microscopy, light *Microscopy* using *light* as the illuminating agent. Often loosely used to include *ultraviolet* and *infra-red microscopy*.

microscopy, ore *Reflected-light microscopy* using *polarized light* for the study (mainly the identification) of ore minerals or other opaque *optically anisotropic* materials.

microscopy, phase-contrast A form of *interference microscopy* (in the widest sense) due to Zernike in which *contrast* in the *image* is enhanced by altering the *optical pathlength* travelled by the *diffracted light* with respect to that travelled by the *direct light*. This is achieved by the action of a *phase plate*. Contrast is achieved by the conversion of *phase differences* in the light leaving the object to *amplitude* differences in the image. Two kinds of phase contrast are available, depending on the characteristics of the phase plate; in positive phase contrast objects which retard the phase of the diffracted light by a small amount appear darker than the background while in negative phase contrast they appear brighter.

microscopy, polarized-light *Microscopy* using *polarized light* in which phenomena due to *optical anisotropy* are made visible and correlated parameters are made measurable.

microscopy, quantitative *Microscopy* in which the phenomena indicating compositional and/or textural elements in the *image* are measured and expressed as optical (physical), *stereological*, and/or geometrical parameters. Quantitative microscopy is a general name for both *microscope photometry* and *image analysis* and may include any kind of microscopy by which quantitative information is obtained.

microscopy, reflected-light *Microscopy* using *illumination* which falls on the *object* from the same side as that from which the object is observed.

microscopy, relief-contrast *Microscopy* in which *relief contrast* is produced in the *image*. This may be achieved by operating with unilateral *oblique illumination* and reducing the intensity of the *direct light* with respect to that of the *diffracted light* by means of an absorbing layer in the form of a strip or ring placed in the *back focal plane* of the *objective* (or in a succeeding plane *conjugate* to this), together with an *illuminating aperture diaphragm* of appropriate shape and size.

Note: The name 'modulation contrast' is used for one method of achieving relief contrast.

microscopy, transmitted-light *Microscopy* using *transmitted-light illumination*.

microscopy, ultraviolet *Microscopy* in which the *image* is formed by *ultraviolet radiation* and is displayed and recorded by means of an *image converter* or photographic film. Microscopy using the near ultraviolet may be performed with a conventional *microscope*; microscopy in the far ultraviolet requires special equipment. Formerly much used to increase the obtainable *resolving power* of the microscope or to observe the presence in unstained *specimens* of certain compounds (such as nucleoproteins) which strongly absorb at specific *wavelengths* in the ultraviolet.

microscopy, video-enhanced contrast *Microscopy* using a *video-enhanced contrast microscope*.

microspectrophotometer A device for *spectrophotometry* of small *specimens* which does not, however, allow visual inspection of the *object* as with a normal *microscope*.

microspectrophotometry The use of a *microspectrophotometer*.

microtome A mechanical device for producing thin *sections* of a *specimen* for study with the *microscope*.

minimum resolvable distance *See* **resolvable distance, minimum**

mired An acronym for MIcroREciprocal Degrees. Used as a means of expressing *colour temperature* (T_c):

$$T_c \text{ in mireds} = \frac{1\,000\,000}{T_c \text{ in kelvin}}.$$

Mired values quoted for *colour correction filters* may be added to the mired value of the *light source* to determine the final colour temperature.

mirror, dichroic *See* **dichroic mirror**

modal analysis [quant.] The procedure by which *modes* are estimated.

mode [quant.] The composition of a heterogeneous material expressed in percentage terms or fractions of volume or weight.

modulation contrast *See* **microscopy, relief-contrast**

monochromat A *lens* in which *chromatic aberration* is minimized for only one *wavelength*. The term is usually used to describe an *objective* made of fused silica and designed to operate at a specific wavelength in the *ultraviolet*.

monochromatic aberrations *See* **aberrations, monochromatic**

monochromatic radiation *See* **radiation, monochromatic**

monochromatism [physiol.] The inability of the *eye* to perceive *colour*.

monochromator A device designed to isolate *monochromatic radiation* from a *complex radiation*.

monocular [physiol.] Using one *eye*.

monocular microscope *See* **microscope, monocular**

monocular tube *See* **tube, monocular**

mount, lens The mechanical parts which support and locate a *lens* within an optical system.

mountant A *mounting medium*.

mounting medium The liquid, synthetic resin or other medium in which the *object* or objects are placed for investigation with the *microscope*. For *transmitted-light microscopy* this medium must be transparent, colourless and of specified *refractive index*, enclosed between the *slide* and the *cover glass*. For *reflected-light microscopy* the mounting medium is normally a resin with which the *sample* is impregnated so that a *polished section* may be made.

movement, Brownian The irregular movement of *microscopic* particles suspended in a fluid caused by the impact on the particles of molecules in the surrounding medium.

multiple-beam interference *See* **interference, multiple-beam**

muscae volitantes [physiol.] *See* **entoptic phenomena**

myopia [physiol.] Nearsightedness. Defective vision in which the *far point* is situated a finite distance in front of the *eye*.

N

narrow-band-pass filter *See* **filter, narrow-band-pass**

natural light *See* **light, natural**

near point (of the eye) [physiol.] The point at which the naked *eye* is focused when it is maximally *accommodated* (i.e. with its highest *refractive power*). Its distance from the eye depends on possible refractive errors of the eye and on the age of the observer.

near-point distance *See* **distance, near-point**

nearest distance of distinct vision The *near-point distance*.

nearsightedness *See* **myopia**

negative eyepiece *See* **lens, diverging**

Nelson, Edward Milles (1851–1938) An English amateur microscopist and astronomer who was responsible for many mechanical and optical improvements to the microscope, and who wrote extensively on microscopy.

* **Nelsonian illumination** *See* **illumination, source-focused**

neutral-density filter *See* **filter, neutral density**

Newton's scale of colours A repeating sequence of colours (divided into 'orders') produced when *white light* passes through a quartz wedge placed between *crossed polars* and oriented at 45 degrees to their axes. The colour sequence may be used in estimating *birefringence* or thickness.

Nicol prism *See* **prism, polarizing**

night vision *See* **vision, night**

nodal plane *See* **plane, nodal**

nodal points *See* **points, nodal**

Nomarski, Georges A physicist working in France who introduced a form of differential interference contrast microscopy.

Nomarski differential interference contrast A form of *differential interference* using special *polarizing prisms* arranged according to a design by Nomarski. It is most frequently applied in *microscopy*.

non-disparate spatial vision *See* **vision, spatial, non-disparate**

normal An imaginary straight line perpendicular to a plane or a surface.

normal position *See* **position, normal**

nosepiece That part of the *body tube* which carries the *objective*.

nosepiece, centring A *nosepiece* equipped with a centring mechanism which allows the position of the *objective* to be adjusted laterally until its *optical axis* coincides with the rotation axis of a *rotating stage*.

nosepiece, revolving A *nosepiece* with a turret which facilitates changing *objectives*.

number, field-of-view *See* **field-of-view number**

numerical aperture *See* **aperture, numerical**

O

object (1) Anything from which an *image* is formed. (2) That which is studied with a *microscope*.

object, auxiliary *See* **plate, retardation**

object, self-luminous An *object* having the properties of a *primary source*. The *light* emitted is *incoherent*; examples are *fluorescent objects* and glowing *filaments* or surfaces.

object distance *See* **distance, object**

object field *See* **field, object**

object finder A precisely made and standardized *slide* carrying a grid of co-ordinates identified by numbers and/or letters, used to specify and relocate an *object field*. The slide bearing the *object* must first be located in a *mechanical stage* or other suitable holder. The finder is substituted for the object slide without moving the stage and the reference co-ordinates noted. Using the same procedure in reverse, the field can be relocated on any *microscope* using the same or another similar finder slide.

object marker *See* **marker, object**

object plane *See* **plane, object**

object-side aperture (of a microscope objective) *See* **aperture, object-side**

object space *See* **space, object**

object to primary image distance *See* **distance, object to primary image**

objective The first part of the imaging system, consisting of a *lens*, its *mount*, and any associated parts. It forms a *primary image* of the object.

objective, achromatic *See* **achromat**

objective, apochromatic *See* **apochromat**

objective, dry An *objective* designed to operate without an *immersion liquid*, i.e. with air between the *front lens* and the *preparation*.

objective, flat-field An *objective* so corrected that the flattening of the *curvature of the image field* in the plane of the *primary image* is emphasized in addition to the *correction* of other *aberrations*. The names for such objectives usually contain the standardized prefix or suffix 'plan', e.g. planachromats or planapochromats.

* **objective, fluorite** *See* **semi-apochromat**

objective, free working distance of The depth of the free space along the *optical axis* between the front of the *objective* and the surface of the *cover glass* or of the *object* itself (if it is uncovered).

objectives, image distance correction of *Objectives* may be corrected for the formation of their *primary image* either at a finite or at an infinite *image distance*. In the former case the objective forms the primary image without the aid of a *tube-lens*, and the level of this image is dictated by the *optical fitting dimensions* of the *microscope*. (See Fig. 2.) In the latter case the objective must be supplemented by a tube-lens in order to form a primary image; whilst the image formed by the objective alone would be at infinity, the image formed with the aid of the tube-lens is in the *back focal plane* of the tube-lens.

objective, infinity-corrected An *objective* corrected for an infinite *object to primary image distance* and which must, therefore, be used with a *tube-lens*. In order to obtain its nominal *magnifying power*, such an objective must be combined with a tube-lens of appropriate *focal length*.

objective, locating flange of (or objective shoulder) The surface of an *objective* which locates it at a given level (that of the *objective-locating surface of the nosepiece*). It is one of the reference planes which determines the *mechanical tubelength* and *parfocalizing distance* (*of objective*). (See Fig. 2.)

objective, long working-distance An *objective* designed to have a longer *free working distance* than a conventional objective of the same *magnifying power*. Often such objectives use a combination of reflecting and refracting elements.

objective, parfocalizing distance of *See* **parfocalizing distance**

objective, reflecting An *objective* which operates by *reflection.* (*See* **microscope, reflecting**).

objective, semi-apochromatic *See* **semi-apochromat**

objective, spring-loaded An *objective* so constructed that the front *lens* and its *mount* (at least) will retract against a spring when brought into contact with the *specimen* or other *object*, thus preventing damage to either object or objective.

objective, strain-free [pol.] An *objective* intended for use with the *polarized-light microscope*, and manufactured by careful selection and mounting of its component parts so as to be entirely free from *double refraction* due to strain.

objective, working distance of *See* **objective, free working distance of**

objective-locating surface (of the nosepiece) The surface of the *nosepiece* which sets the level of the *locating flange* of the *objective*. It is one of the *reference planes* which determine the *mechanical tubelength*, the *parfocalizing distance (of objective)*, and the *primary-image distance (of the objective)*. (See Fig. 2.)

objective shoulder *See* **objective, locating flange of**

objective to primary image distance *See* **distance, objective to primary image**

oblique extinction *See* **extinction, oblique**

oblique illumination *See* **illumination, unilateral oblique**

observation tube *See* **tube, viewing**

ocular *See* **eyepiece**

ocular, slotted *See* **eyepiece, slotted**

ocular dominance *See* **dominance, ocular**

oil, immersion *See* **immersion oil**

oil-immersion lens *See* **lens, immersion**

opal glass *See* **glass, opal**

optic axis *See* **axis, optic**

optical activity *See* **activity, optical**

optical anisotropy *See* **anisotropy, optical**

optic axial angle *See* **angle, optic axial**

optical axis *See* **axis, optical**

optical density (internal transmission density or absorbance) *See* **absorbance**

optical fitting dimensions *See* **dimensions, optical fitting (of the microscope)**

optical indicatrix *See* **indicatrix**

optical path *See* **path, optical**

optical pathlength *See* **pathlength, optical (optical distance)**

optical pathlength difference *See* **pathlength difference, optical**

optical rotation *See* **rotation, optical**

optical thickness *See* **pathlength, optical (optical distance)**

optical tubelength *See* **tubelength, optical**

optically biaxial *See* **biaxial, optically**

optically uniaxial *See* **uniaxial, optically**

optics, geometrical A form of representation of propagation of *radiation* under the following conditions: it is along a straight line; *beams* emanating from a body are independent of one another; the laws of *refraction* and *reflection* are adhered to. Geometrical optics is illustrated using *rays*, and does not take into consideration the wave nature of radiation, nor the phenomena of *interference, coherence,* and *diffraction*. In geometrical optics the *object* points are imaged as *image* points at the intersection of rays rather than as *diffraction discs (Airy discs)* as in *wave optics*.

optics, wave That branch of optics which explains *image* formation in terms of the wave theory of *radiation*, and deals with the phenomena of *interference, diffraction,* and *polarization*.

ordinary ray *See* **ray, ordinary**

ore microscope *See* **microscope, ore**

ore microscopy *See* **microscopy, ore**

organ, visual [physiol.] The functional union of the *eyes* and those parts of the brain which contribute to the visual process.

orthoscopic eyepiece *See* **eyepiece, Kellner**

orthoscopic observation Observation of the normal *image* in a *microscope*.

overcorrection An error in the *correction* of *spherical aberration*, leading to an unsharp *image,* which lacks in *contrast.* In *microscopy* it may be caused by the use of a *cover glass* thicker than, or a *mechanical tubelength* longer than, the values assumed in the computation of the *objective*.

P

pancratic Signifies that the *focal length* of an optical system can be changed continuously between limits by moving a single *lens* or group of lenses. A pancratic system is adjustable to different degrees of *magnification* without altering the positions of the *object* or *image planes*. This classical name is becoming replaced superfluously by the word 'zoom'. (*See also* **condenser, pancratic**)

paraboloid condenser *See* **condenser, paraboloid**

parallax [physiol.] The apparent lateral displacement with respect to one another of two *objects* situated at different axial distances with lateral change in the direction of observation. Parallax can contribute to the perception of depth.

parallel polars *See* **polars, parallel**

parallel ray bundle *See* **bundle, ray, parallel**

parallel ray path *See* **ray path, parallel**

paraxial Pertaining to the *Gaussian space*.

paraxial ray *See* **ray, paraxial**

parfocal Signifies that once any *lens* of a set (e.g. *objective, tube-lens* or *eyepiece*) has been *focused* on to the *object* or adjusted so that the *image* lies at its correct level, then if it is exchanged for any others in that set at a constant setting of the *microscope*, no readjustment of *focus* will be necessary to restore maximal sharpness. A small readjustment may be needed, however, because of the *accommodation* which may take place in the *eyes* of an observer.

parfocalizing distance (of the eyepiece) The distance between the *primary image plane* and the *eyepiece-locating surface* (of the viewing tube). It is one of the *optical fitting dimensions* and is now commonly 10 mm. (See Fig. 2 and Table 2.)

parfocalizing distance (of the objective) The distance in air between the *object plane* (i.e. the uncovered surface of the *object*) and the *locating flange of the objective,* when the *microscope* is in its working position. It is one of the *optical fitting dimensions*, now commonly 45 mm in air. (See Fig. 2 and Table 2.)

partial coherence *See* **coherence, partial**

patch stop *See* **darkground stop**

path, light The space through which *light* passes from its *source* to the final sensing element in the system according to the rules of *geometrical optics*, i.e. neglecting *diffraction* effects. Its extent is controlled by the diameter of *field* or *aperture diaphragms*.

path, optical A path in a given direction through which *light* passes.

path, ray *See* **ray path**

pathlength, optical (optical distance) In a homogeneous *medium* this is the product of the geometrical length of the optical path and the *refractive index* of the medium containing that path. It is expressed either in length units or as a fraction or multiple of a given *wavelength*. When the medium is inhomogeneous it is the sum or integral of the product of the geometrical lengths and refractive indices of the parts.

pathlength difference, optical The difference (expressed in length units or *wavelengths*) in *optical pathlength* between two optical paths due to differences in geometrical length, *refractive index*, or both. The term is usually used to denote the difference between two *coherent wavetrains* which may *interfere*.

peak wavelength *See* **wavelength, peak**

pencil, ray A term used in *geometrical optics* instead of the term *'bundle'* when the cross section of the assembly of *rays* is almost linear. As with *ray bundles,* ray pencils may be convergent, divergent, or parallel.

phase Relative position in a cyclical or wave motion. It is expressed as an angle, one cycle or one *wavelength* corresponding to 2π radians or 360 degrees.

Note: The term 'in phase' corresponds to phase angles between two occurrences of 0 or 2π radians (360 degrees) or a whole number multiple of these.

phase-contrast microscopy *See* **microscopy, phase-contrast**

phase difference The *phase angle* or fraction or number of *wavelengths* by which one periodic disturbance or wave lags behind or precedes another in time and space.

phase plate An optical device used in *phase-contrast microscopy*. It is placed in the *back focal plane* of the *objective* (or in the plane of a succeeding *image* of it), where it receives an image of a *diaphragm* (usually annular) positioned in the *front focal plane* of the *condenser*. It influences differently the *phase* and *amplitude* of the direct and diffracted light.

phase-shift An alteration in the *phase* introduced by the interaction of a wave with matter.

phase velocity The velocity with which a locus of a *wave group*, in which all waves of different length have equal *phase* (e.g. the phase at which maximum *amplitude* occurs), travels in a given spatial direction.

phenomena, entoptic *See* **entoptic phenomena**

phosphorescence A form of *photoluminescence* which persists for an appreciable time (usually more than 10^{-8} s) after the cessation of the *excitation*.

photoluminescence The phenomenon of selective absorption of *radiation* of relatively short *wavelength* (i.e. of high energy) by matter, resulting in the emission of *radiation* of longer wavelengths (i.e. of lower energy).

photomacrography The production of a photographic *image* of an *object* with a reproduction ratio on the negative from approximately 1:1 to about 15:1.

photometer An instrument for measuring *light* in terms of *photometric quantities*. Originally this was done by comparison against a standard using the *eye*, but it is now performed by physical instruments more properly termed *radiometers*.

photometer head *See* **head, photometer**

photometer tube *See* **tube, photometer**

photometric Pertaining to a *photometer* or to *photometry*.

photometric contrast *See* **contrast, photometric**

photometric diaphragm *See* **diaphragm, photometric**

photometric field *See* **field, photometric**

photometric quantity A quantity which evaluates the *intensity* of *light* according to the *maximum luminous efficacy* (K_{max}) and the *relative spectral luminous efficiency* ($V(\lambda)$) of the eye. With *monochromatic radiation* it is related to the corresponding *radiometric quantity* by the expression

$$dX_v(\lambda) = K_{max}\, dX_e(\lambda)\, V(\lambda)$$

where X_v and X_e are one of the photometric quantities and one of the corresponding *radiometric quantities* (in a *wavelength interval* of differential width) listed in Table 1; (λ) indicates the central wavelength of monochromatic radiation. In practice, the differential width is replaced by a narrow finite width $\Delta\lambda$ and dX by ΔX. With radiation extended across the total visible *spectrum* the relation between photometric and radiometric quantities is

$$X_v = K_{max} \int_{\lambda=380}^{\lambda=780} X_{e,\lambda}\, V(\lambda)\, d\lambda$$

where the subscript λ indicates *spectral concentration*.

photometry The measurement of *radiation* or effects related to the interaction of *light* with matter assessed visually and expressed in *photometric quantities*. Now objectively evaluated by *radiometry*. (See Table 1.)

photometry, microscope The application of *photometry* to the investigation of *microscopic objects* which are simultaneously observed with the *microscope*.

Note: In this context the word photometry is used although the measurement generally involves radiometric quantities.

photomicrograph A photographic record of an *image* formed by a *microscope*.

photomicrographic field diaphragm *See* **diaphragm**

photomicrographic magnification *See* **magnification, photomicrographic**

photomicrography The recording by photography of an *image* formed by a *microscope*; i.e. photography through a microscope.

Note: Not to be confused with *microphotography*.

photon Elementary quantity of *radiant energy* (quantum) whose value is equal to the product of Planck's constant *h* and the frequency (in hertz) of the electromagnetic radiation.

photopic vision *See* **vision, daylight**

physiological contrast *See* **contrast, physiological**

pincushion distortion *See* **distortion, pincushion**

plan-objective (or plano-objective) *See* **objective, flat-field**

plane, aperture That plane in which an inserted *diaphragm* will act as an *aperture diaphragm*, and any plane *conjugate* with it.

plane, field The *object plane* and any plane *conjugate* with it. A *diaphragm* inserted in a field plane will act as a *field diaphragm*.

plane, focal *See* **focal plane**

plane, image Any *field plane* in which an *image* is situated.

plane, image, primary The *image plane* in which the *primary image* is formed. The primary image plane is important as one of the *reference planes* for the *optical fitting dimensions*. (See Figs. 2, 8, and 9.)

plane, nodal The two *cardinal planes* of a *lens*, which (in the *paraxial* space) are intersected by a *ray* (R_k) entering or leaving the lens at the same angle to the *optical axis*, i.e. parallel with each other (see Fig. 5). If the *refractive indices* of the media bounding the lens at the object side (n) and the image side (n') are equal, nodal planes and *principal planes* coincide. The intersections of nodal planes with the optical axis are *nodal points*.

plane, object That *field plane* in which the *object* is situated. The object plane is important as one of the *reference planes* for the *optical fitting dimensions*. (See Figs. 2 and 4.)

plane, polarization The plane containing the *polarizing direction* and the direction of wave propagation.

plane, principal For a hypothetical infinitely *thin lens*, this is the plane perpendicular to the *optical axis* and coinciding with the lens, from which both object-side and image-side *focal lengths* are measured. A practical *thick lens* has two principal planes separated by a distance along the optical axis (the *hiatus*) which is characteristic for each system: the object-side principal plane from which the object-side focal length (f) and the *object distance* (a) are measured; and the image-side principal plane from which the image-side focal length (f') and the *image distance* (a') are measured. The object-side principal plane (H) is usually situated along the optical axis nearer to the object than the image-side principal plane (H'). The points where the principal planes intersect the optical axis are known as the *principal points*. (See Figs. 4 and 5.)

plane, reference, for optical fitting dimensions *See* **dimensions, optical fitting, reference plane for**

plane, vibration (of electromagnetic radiation) The plane containing the *vibration direction* and the direction of wave propagation.

plane-polarized light *See* **light, plane-polarized**

planes, cardinal In *geometrical optics* the reference planes erected normal to the *optical axis* passing through the *cardinal points* of a *lens* or an optical system. Used when considering the imaging function and for the representation of *rays* in diagrams. The cardinal planes are: *focal planes; principal planes; nodal planes;* the *object plane;* and the *image plane*. (See Fig. 4.)

planes, conjugate *See* **conjugate planes**

plate, first-order red [pol.] A *retardation plate* producing the *interference colour* having the typical tint of the *first-order red*. Sometimes called a 'sensitive tint plate'.

plate, half-wave [pol.] A *retardation plate* producing an *optical pathlength difference* of half a *wavelength*, the reference wavelength being as a rule taken to be about 550 nm.

plate, quarter-wave [pol.] A *retardation plate* producing an *optical pathlength difference* of a quarter of a *wavelength*; the reference wavelength is selected according to the application and is individually indicated. It changes *plane-polarized light* into *circularly polarized light*.

plate, retardation [pol.] A piece, or pieces, of *optically anisotropic* material with plane faces inserted between the crossed *polars* in a *diagonal position* to produce a specific *optical pathlength difference* of mutually perpendicular *plane-polarized light* waves. Such plates are sometimes called 'auxiliary objects'.

plate, zone A transparent flat plate divided into a series of zones by circles whose radii are in the ratios of $1:\sqrt{2}$, $1:\sqrt{3}$, $1:\sqrt{4}$, etc. Alternate zones are blackened. A plane *light* wave incident normally on the plate is diffracted so that light is concentrated at a certain point on the *optical axis* of the plate, which thus acts as a *lens*.

pleochroism [pol.] The property of an *optically anisotropic* medium by which it exhibits different *brightness* and/or *colour* in different directions of *light* propagation or in different *vibration directions* on account of variation in selective *spectral absorption* of transmitted light.

pleochroism, reflection *See* **bireflection**

point, convergence [physiol.] That point in the *object space* at which the visual axes of both *eyes* intersect in normal *binocular* vision.

point, fixation [physiol.] The point towards which a viewing *eye* is directed and aligned. It represents the object-side end of the *fixation line*.

point counter [quant.] A *mechanical stage* designed or additionally equipped to move the *object* in steps in order to facilitate *modal analysis* and the evaluation of stereological parameters.

point source (of light) *See* **source, point**

points, cardinal In *geometrical optics* the reference points on the axis of a *lens* used when considering the imaging function and for the representation of the *rays* in diagrams. They are: the *focal points* (F on the *object side* and F′ on the *image side*); the *principal points* (H on the object side and H′ on the image side); the *nodal points* (K on the object side and K′ on the image side); the *object point* (O); and the *image point (O′). (See Figs. 4 and 5.)*

points, conjugate Those points in both *object space* and *image space* which are imaged one on to the other.

points, nodal The intersections of the *nodal planes* with the *optical axis*. (See Fig. 5.)

points, principal The two *cardinal points* (H and H′) of a *lens* where the *optical axis* is intersected by the *principal planes*. The principal points are the reference points from which the *focal lengths* (f and f'), the *object distance* (a), and the image distance (a') are measured. (See Figs. 4 and 5.)

polar A term introduced in 1948 by A. F. Hallimond for any device which selects *plane-polarized light* from *natural light.*

polarization, degree of The proportion of *plane-polarized light* to total light in a mixture with *natural light.*

polarization direction, (of electromagnetic radiation) The direction of the magnetic vector of electromagnetic *radiation*; it is perpendicular to the *vibration direction* and the direction of wave propagation.

polarization figure *See* **figure, polarization**

polarization plane *See* **plane, polarization**

polarization state The mode of polarization such as linear, elliptical, or circular and the orientation of planes related to polarization.

polarized light *See* **light, polarized**

polarized-light microscope *See* **microscope, polarized-light**

polarized-light microscopy *See* **microscopy, polarized-light**

polarizer A *polar* placed in the *light path* before the *object.*

polarizing filter *See* **filter, polarizing**

polarizing interference *See* **interference, polarizing**

polarizing prism *See* **prism, polarizing**

Polaroid A trade name for a *polar* in the form of a thin plastic sheet. Invented by Dr E. H. Land in 1932

Polaroid film A type of photographic film manufactured by the Polaroid Land Corporation and supplied in a pack with the processing chemicals, so enabling the final *image* to be produced without a darkroom and within a few minutes of making the exposure.

polars, crossed The condition in which the *vibration directions* of *polars (polarizer* and *analyser)* are mutually perpendicular.

polars, parallel The condition in which the *vibration directions* of the *polars (polarizer* and *analyser)* are parallel.

polars, uncrossed The condition in which either the *polarizer* or *analyser* is slightly rotated (with respect to the perfect *crossed* position) by an angle which may be specified.

polished section A *preparation* made for study with a *reflected-light microscope.* It is cut from a compact *sample* or taken from a mass of hardened resin which contains a single *object* or bulk objects. The main property of a polished section is that it has at least one surface for investigation which has been ground and then polished and is as flat as possible so that it behaves like a flat mirror. The surface may be coated or etched in order to enhance *contrast* between components of the specimen. (*See also* **levelling)**

portable microscope *See* **microscope, portable**

position, addition [pol.] The state of mutual orientation of two transparent *double-refracting media* so that their *vibration directions* are parallel and the vibration direction relating to the higher *refractive index* in one medium coincides with the higher refractive index in the other medium. The total *optical pathlength difference* of both media is then obtained by the addition of the optical pathlength differences of the media.

position, diagonal [pol.] The state of orientation of a *double-refracting object* so that its *vibration directions* form angles of 45 degrees with the vibration directions of the *crossed polars.*

position, extinction [pol.] The state of orientation of a *double-refracting object* so that it shows *extinction* between *crossed polars.*

position, normal [pol.] The state of orientation of a *double-refracting object* so that its *vibration directions* are parallel with those of the *crossed polars.*

position, subtraction [pol.] The state of mutual orientation of two transparent *double-refracting media* so that their *vibration directions* are parallel and the vibration direction relating to the higher *refractive index* in one medium coincides with the lower refractive index in the other medium. The total *optical pathlength difference* of both media is then obtained by subtraction of the optical pathlength difference of one medium from that of the other.

* **positive eyepiece** *See* **eyepiece, positive**

positive lens *See* **lens, converging**

power (1) The ability to do work. (2) The rate of doing work (the SI unit is the watt (W)).

power, collecting (of a lens) *See* **collecting power of a lens**

power, dispersive *See* **dispersive power**

power, magnifying *See* **magnifying power**

power, refractive *See* **refractive power**

power, resolving *See* **resolving power**

preparation (1) The *specimen* or *sample* (or any part of it) as made up for study with the microscope. (2) The act of making the specimen ready for study with the microscope.

presbyopia [physiol.] The reduction of the *accommodation* power of the *eye* by ageing processes, especially involving a reduction in the elasticity of the *lens* of the eye. The term is usually applied if the range of *accommodation* is reduced below about 4 dioptres.

primary diffraction pattern (or image) *See* **diffraction pattern (or image), primary**

primary image The (usually) magnified *real image* of the *object* formed by the *objective* at a place determined by the *optical fitting dimensions*. (See Figs. 2, 8, and 9.)

primary image distance (of the objective) *See* **distance, objective to primary image**

primary image plane *See* **plane, image, primary**

primary interference image *See* **interference image, primary**

primary magnification *See* **magnifying power, of objective**

primary source *See* **source, primary**

principal plane *See* **plane, principal**

principal point *See* **point, principal**

principal ray *See* **ray, principal**

principal vibration directions *See* **directions, vibration, principal**

prism A block of transparent refractive material limited by at least two intersecting planes, used to disperse *light* or deviate it through an angle.

prism, drawing *See* **drawing prism**

prism, Nicol A type of *polarizing prism*.

Note: The name 'Nicol' to indicate a polar in general should not be used.

prism, polarizing Two pieces of *double-refracting* material, e.g. calcite or quartz, or one of these plus a piece of glass, cemented together to form a double *prism* which acts by *refraction* and *total internal reflection* or by refraction only. It splits a *beam* of *natural light* into two beams of *plane-polarized light* having mutually perpendicular *vibration directions* and being propagated in two different directions. When one of these beams is removed, e.g. by *absorption*, the prism acts as a *polar*, otherwise it may be used as a *beam splitter*. Many types of polarizing prism exist, most known by the name of their originator, e.g. Nicol, Wollaston.

prism, Wollaston A double-image *polarizing prism* made of two geometrically similar, wedge-shaped *prisms* of quartz or calcite cemented together. The cutting of the two prisms is arranged so that their *crystal optical axes* produce two coloured, oppositely polarized emergent *light beams*. It is useful in determining the degree of *polarization* in a partially polarized *beam* or as a *beam splitter* for differential interference, providing for examination two *images* with vibrations in two perpendicular directions.

projection eyepiece *See* **eyepiece, projection**

projection lens *See* **lens, projection**

pseudoscopy A reversal of the sensation of depth due to the exchange of the *images* intended for the right and left *eyes* in a stereoscopic system.

pupil The apparent minimum common cross section of all *ray bundles* both on the *object side* (the entrance pupil) and on the *image side* (the exit pupil).

Note: It may indicate an aperture or the image of an aperture. The centre of the pupil is the point of intersection of principal rays with the optical axis.

pupil, entrance *See* **pupil**

pupil, entrance, of the eye [physiol.] The object-side *image* of the *iris* of the *eye*, which works as the *aperture diaphragm*. In the normal eye (an eye with its optical values standardized on the basis of statistical studies), the *entrance pupil* is located 3 mm behind the *vertex* of the *cornea*.

pupil, exit *See* **pupil**

pupil of the eye [physiol.] The variable *aperture* in the *iris* of the *eye* through which *light* enters.

purity [photom.] (1) A measure of the *saturation of colour*. It is the ratio between the distance on a *chromaticity diagram* from the *chromaticity co-ordinates* of a standard *source* to those of the *sample*, and the distance between the chromaticity co-ordinates of the source and those of the intersection between a straight line joining the co-ordinates of the source and the sample with either the *spectrum locus* or the *purple line* (see Fig. 10). (2) The proportion of the pure monchromatic component in a spectral mixture.

Note: Purity may be expressed as a fraction or as a percentage.

Purkinje, Jan E. (1787–1869) A Czech physiologist and researcher in experimental physiology and psychology, histology and embryology.

Purkinje shift [physiol.] The movement of the point of maximum sensitivity of the human *eye* towards the blue end of the *spectrum* at low *light* levels.

purple line [photom.] A straight line on the *chromaticity diagram* joining the extreme ends of the *spectrum locus* . (See Fig. 10.)

Q

quantitative microscopy *See* **microscopy, quantitative**

quantity (of light) [photom.] Product of *luminous flux* and its duration. (See Table 1.)

quantity, radiometric *See* **radiometric quantity**

quantum efficiency The number of electrons released in a *photon* detector per photon of incident *radiation* of specified *wavelength*.

quarter-wave plate *See* **plate, quarter-wave**

quartz wedge *See* **wedge, quartz**

Quekett, John Thomas, (1815–61) A physician, later Professor of Histology at the Royal College of Surgeons. Author of a famous treatise on the use of the microscope. Honorary Secretary of the Microscopical Society of London 1842–59 and President from 1860 until his death.

R

radiance [photom.] *Radiant flux* leaving, arriving at, or passing through a unit area of surface. (See Table 1).

radiant energy *See* **energy, radiant**

radiant exitance *See* **exitance, radiant**

radiant exposure *See* **exposure, radiant**

radiant flux *See* **flux, radiant**

radiant intensity *See* **intensity, radiant**

radiation (1) [photom.] The *emission* or transfer of energy in the form of electromagnetic waves or particles. (2) The waves or particles themselves.

Note: The radiation may be qualified as electromagnetic, unless this is clear from the context.

radiation, black-body [photom.] The quality and quantity of the *radiation* (depending solely on its absolute temperature) which is emitted by a *black-body radiator*.

Note: The radiation is described mathematically by Planck's radiation formula.

radiation, complex *Radiation* consisting of a mixture of *monochromatic radiations*.

radiation, infra-red Electromagnetic *radiation* in which the *wavelengths* of its *monochromatic* components are longer than those for *light* and less than about 1 mm.

Note: Microscopy using radiation with wavelengths up to about 1100 nm (which is known as 'near infra-red') may also be carried out since many normal microscopes permit work with this radiation using an image converter for observation.

radiation, monochromatic *Radiation* consisting of only a single *wavelength*, or of only a very narrow band of wavelengths, in which case the central wavelength is quoted.

radiation, ultraviolet Electromagnetic *radiation* in which the *wavelengths* of its *monochromatic* components are shorter than those of *light* and longer than about 100 nm.

Note: In general microscopy the shortest usable wavelength is about 380 nm. When using ultraviolet transmitting or reflecting optical systems the spectral working range can be extended down to about 240 nm.

radiation, visible *See* **light**

radiator, black-body [photom.] A *black-body* which is heated, thus becoming an emitter of *radiation*.

radiometer [photom.] An instrument for measuring *radiation* in terms of *radiometric quantities*.

radiometric [photom.] Related to the measurement of *radiation*.

radiometric quantity [photom.] A physical quantity which evaluates the strength of *radiation*, e.g. power (SI unit watt), energy, or work (SI unit joule), or *areal, spatial,* or *spectral* densities (concentration) of these parameters (see Table 1). It may be converted into a photometric quantity.

radiometry [photom.] The measurement of *radiation* or effects related to its interaction with matter, expressed as *radiometric quantities*.

Ramsden, Jesse, (1735–1800) An English instrument-maker who developed achromatic telescopes and a micrometer microscope for reading the fine circle divisions on instruments.

Ramsden disc *See* **microscope, exit pupil of**

Ramsden eyepiece *See* **eyepiece, Ramsden**

raster A pattern of lines or points systematically applied to an area, in a (usually) regular or random manner.

ray A geometrical abstraction; a straight line which describes a direction of propagation of *radiation*, and which in an isotropic *medium* is normal to the propagated wavefront. It may be used to designate the limits and/or axes of *ray bundles, beams*, or a longitudinal section through a *light path*.

ray, axial A *ray* which passes along the *optical axis*.

ray, conjugate portions of Those portions or sections of a *ray* in both *object space* and *image space* which are imaged one on to the other.

ray, extraordinary That one of the two *rays* transmitted through a *double refracting medium* which travels with a velocity which varies with the direction of transmission.

ray, incident A *ray* which intercepts a surface.

ray, marginal A *ray* joining the centre of a *field* to a point at the edge or margin of the next *pupil* or *aperture*; thus the centre of a field is the point of intersection of all marginal rays.

ray, ordinary That one of the two *rays* transmitted through a *double refracting medium* which travels with a velocity which does not vary with direction.

ray, paraxial A *ray* which is almost parallel to the *optical axis* and contained in the *Gaussian space*. For paraxial rays it may be assumed that sin $\sigma = \sigma$ when making calculations involving these rays.

ray, principal A *ray* joining any point in a *field* to the centre of the next *pupil* or *aperture*; thus the centre of a pupil or aperture is the point of intersection of all principal rays. (See Fig. 4.)

ray bundle *See* **bundle, ray**

ray path The graphical representation of a *light path* by means of selected *rays*, e.g. *principal*, and/or *marginal*, and/or *axial* rays.

ray path, parallel A *ray path* in which all *rays* originating from individual points in the anterior *focal plane* of a *converging lens* are parallel with each other and thus form parallel *ray bundles*.

Note: In a parallel ray path the individual ray bundles are not parallel with each other. The parallel ray path is used between two converging lenses to avoid the influence of changes in distance or optical pathlength between the lenses on the quality or size of the image, or its position with respect to the nearest lens.

ray pencil *See* **pencil, ray**

ray space *See* **space, ray**

Rayleigh, Lord (formerly William Strutt) (1842–1919) A physicist who worked in the fields of physical optics, the resolving power of optical instruments and the wave theory of sound.

real image *See* **image, real**

red, first-order (sensitive tint) [pol.] The characteristic reddish-violet *interference colour* selected from *light* of a *continuous spectrum* by the *extinction* of the adjacent *wavelengths*. (*See* **plate, first-order red**)

reference directions *See* **directions, reference**

reference plane (for optical fitting dimensions) *See* **dimensions, optical fitting, reference plane for**

reference viewing distance An internationally agreed standardized distance of 250 mm between an *object* and the vertex of the *cornea* of the *eye*. This term supersedes the older 'nearest distance of distinct vision' in optical calculations.

reflectance [photom.] The measure of *reflection* in terms of the ratio (expressed as a fraction or as a percentage) of the reflected *intensity* to the incident intensity.

reflected-light illumination *See* **illumination, epi-**

reflected-light illuminator *See* **illuminator, epi-**

reflected-light microscope *See* **microscope, reflected-light**

reflected-light microscopy *See* **microscopy, epi-**

reflection Return of *radiation* from a surface without change of proportions of the *monochromatic* components of which the *radiation* is composed.

reflection, angle of The angle between the *normal* to a surface and a *ray* reflected from it.

Note: For specular reflection this angle is equal to the angle of incidence; for diffuse reflection this need not be so.

reflection, diffuse Diffusion by *reflection* in which, on the macroscopic scale, there is no *specular reflection*.

reflection, specular (reflection, regular) *Reflection* without *diffusion* in accordance with the laws of optical reflection, as by a mirror.

reflection, total internal *Reflection* at the interface between two media, when *radiation* travels in a medium of high *refractive index* towards one of lower refractive index. It occurs when the *angle of incidence* on the interface is greater than the *critical angle*.

reflection pleochroism *See* **bireflection**

reflection rotation [pol.] The rotation of the *vibration plane* of *plane-polarized light* due to *bireflection* (if the *object* is in the *diagonal position*).

reflector The optical element in a *reflected-light microscope* (e.g. glass plate, prism, or ring-shaped mirror) which reflects the *light* from the axis of the *illuminator* into the *microscope* axis. The design of the reflector depends on the method of *illumination*.

refraction A change in the direction of propagation of *radiation* determined by the change in the velocity of propagation on passing through an optically non-homogeneous medium or on passing from one medium to another in a direction other than the *normal* to the interface.

refraction, double *See* **double refraction**

refraction, of the eye [physiol.] The refractive condition of the individual *eye*.

refraction increment, specific The increase in *refractive index* of a solution for every 1% increase in solute concentration. Symbol α.

refractive index The ratio of the speed of *light* (more exactly, the *phase velocity*) in a vacuum to that in a given medium (symbolized by the letter n or n').

refractive index, principal One of the *refractive indices* due to *double refraction* and related to *light* vibrating in one of the *principal vibration directions*, by which it is specified, using the symbol of the vibration direction as a subscript to the refractive index symbol, e.g. n_o, n_e, n_x, n_y, n_z, or n_ϵ, n_ω, n_α, n_β, n_γ.

refractive power (1) The ability of a transparent body to refract *light*. (2) The reciprocal of the *focal length* of a *lens* (unit is the dioptre).

refractometry The measurement of *refractive index*.

refractometry, immersion A method for estimating the *refractive indices* of transparent *objects* by immersing them in *media* of varying refractive index. The refractive index of the object is that of the medium in which the minimum *contrast* is observed.

relative dispersion *See* **dispersive power**

relay lens *See* **lens, relay**

relief (1) The differences in height of a surface, e.g. of a flat sculpture. When illuminated from one side, such an *object* shows a characteristic distribution of *light* and shadow which enables the observer to recognize the three-dimensional form of the object. (2) The light distribution appearing in *microscopy* using azimuthal methods (e.g. unilateral *oblique illumination, relief contrast, differential interference contrast*) at the interfaces of object elements with different *optical pathlength*, even when no geometrical relief exists, may also be described as relief. In this case the light distribution appears similar to that produced by a genuine relief. Care must be taken to avoid misinterpreting *optical pathlength differences* as geometrical ones. A further misinterpretation of relief may be caused by *inversion*.

relief contrast A form of *contrast* which presents gradients of geometrical or *optical pathlength differences* in the *object* in the form of a distribution of *brightness* in the *image* which gives an impression of *relief*. This impression occurs because the distribution of brightness in a relief contrast image is similar to the distribution of light and shadow in the image of a three-dimensional object illuminated from one side.

relief-contrast microscopy *See* **microscopy, relief-contrast**

resolution[1] The act or result of displaying fine detail in an *image*. Sometimes used loosely to refer to its quantitative expression, the *resolved distance*.

resolvable distance, minimum[1] The smallest separation of points in an *object* which can be recognized as distinct in an *image*. In *microscopy* this is normally expressed in units of length (μm or nm).

resolved distance[1] A distance equal to or greater than the *minimum resolvable distance*. It is that distance which is resolved

[1] When used without any qualification, the above terms refer to distances at right angles to the *optical axis*.

in practice under any given set of conditions.

resolving power[1] The ability to make points or lines which are closely adjacent in an *object* distinguishable in an *image*. High resolving power implies that the resolved distance is small.

resolving power, diffraction-limited *See* **diffraction limit (of resolving power)**

retardation The slower propagation of a wavefront in a *medium* of high *refractive index* as compared with that in a medium of low refractive index.

retardation plate *See* **plate, retardation**

reticle A *graticule* with a pattern in the form of a square grid.

retina [physiol.] The light-sensitive layer lining the posterior aspect of the inner surface of the *eye* upon which the *image* is formed. It is a multi-layer structure containing photoreceptors (*rods* and *cones*) and other nervous elements involved in processing the image. Its characteristics set one of the limits to the *resolving power* of the eye.

retina, cones of *See* **cones (of the retina)**

retina, rods of *See* **rods (of the retina)**

retinal image *See* **image, retinal**

revolving nosepiece *See* **nosepiece, revolving**

Rheinberg illumination *See* **illumination, Rheinberg**

ring light *See* **light, ring**

rods (of the retina) [physiol.] One of the two types of sensory cell in the *retina*. They serve for perception in *twilight (mesopic) vision* and in *night (scotopic) vision*. They do not contribute to colour sensation and contain the pigment rhodopsin (visual purple).

RMS thread *See* **screw thread, for objective**

rotating stage *See* **stage, rotating**

[1] When used without any qualification, the above terms refer to distances at right angles to the *optical axis*.

rotation, optical [pol.] The rotation of the *vibration direction* of *plane-polarized light* around the axis of *light* propagation in a medium due to *optical activity*. This phenomenon results in a twisting of the *vibration direction* into a helicoidal surface.

Note: Optical rotation occurs in two senses: laevo rotation which takes place in an anticlockwise direction to an observer facing the oncoming light, and dextro rotation which is in the opposite (clockwise) sense.

rotation, specific [pol.] A short name for *optical rotation* calculated for a *light path* of 1 mm in length.

Note: Specific rotation is a function of wavelength; in transparent media it increases (like refractive index) with decrease in the wavelength.

S

sample (1) A representative small portion to show the quality of the whole. (2) That which is to be studied with the *microscope*.

saturation of colour [physiol.] A subjective assessment of the degree to which a *colour* departs from white and approaches the pure colour of a *spectral* line. (*See* **purity**)

scale bar A line of calculated length drawn on a *micrograph* to indicate the length in the micrograph of a stated length in the *object*.

scale, imaging *See* **magnification, lateral**

scanning microscope *See* **microscope, scanning optical**

scanning stage *See* **stage, scanning**

scotopic vision *See* **vision, night**

screen A reflecting or translucent surface on which a *real image* may be formed and observed.

screen, diffusing A translucent *screen* used to break up the structure of an unwanted *image*, e.g. that of a *lamp filament*.

screw micrometer *See* **eyepiece, micrometer-screw**

screw thread, for objective A screw thread for connecting a *microscope objective* to the *nosepiece*. Originally standardized by the Royal Microscopical Society and now internationally defined by the ISO Standard 8038. (See Table 2.)

secondary source *See* **source, secondary**

self-luminous object *See* **object, self-luminous**

semi-apochromat An *objective* intermediate in terms of its *correction* and complexity of construction between *achromats* and *apochromats*.

Senarmont, Henri H. de (1808–62) A French mineralogist and crystallographer who introduced the quarter-wave plate in 1840.

sensitive-tint plate *See* **red, first-order**

shearing interference *See* **interference, shearing**

short-wave-pass filter *See* **filter, short-wave-pass**

sign of birefringence *See* **birefringence, sign of**

simple microscope *See* **microscope, simple**

sine condition A mathematical formulation for the condition under which an optical system free from *spherical aberration* is also free from *coma,* i.e. is *aplanatic*. It may be stated as

$$\frac{n \sin \sigma}{n' \sin \sigma'} = M_1$$

where n and n' are the *refractive indices* of media in the *object* and *image* space respectively, and σ and σ' are the *angular apertures* on the object side and the image side respectively. M is the *lateral magnification* of the image and must be a constant for all rays contributing to the image (See Fig. 6.)

slide (1) A flat rectangular plate of glass on which an *object* is mounted for *microscopical* examination. For calculation and *correction* of the *condenser* it is regarded as part of the *condenser*, so that its thickness, *refractive index* and *dispersive power* must be adapted to the demands of the condenser. These parameters together with its length and width are defined by the ISO Standard 8037/1 (see Table 2). (2) A longitudinal mechanical bearing surface upon which parts of the *microscope* move relative to each other.

slide, hanging-drop A *microscope slide* provided with a concavity to accommodate a *hanging drop*.

sliding stage *See* **stage, sliding**

slotted eyepiece *See* **eyepiece, slotted**

Snell, Willebrord van R. (1580–1626) A Dutch astronomer and mathematician who in 1621 discovered the law of refraction which now bears his name.

Snell's law The basic law of *refraction*. It is expressed as

$$\frac{\sin i}{\sin r} = \frac{n_2}{n_1}$$

where i is the angle between the *incident ray* and the *normal* to the surface, r is the angle between the refracted ray and the normal, and n_1 and n_2 are the *refractive indices* of the *media* on the incident and refracted sides of the interface respectively.

* **society thread** An old term for the *screw thread* for the *objective*.

solid angle [photom.] The included angle at the apex of a cone. The unit of measurement is the steradian (sr) which is defined as the solid angle subtended at the centre of a sphere by an area on the surface equal to the radius squared. The solid angle is an important quantity for *photometric* or *radiometric* calculations.

Sorby, Henry Clifton (1826–1908) An English scientific amateur who became a full-time research scientist studying geology and metallurgy. He was one of the first to use the microscope for the study of metals. He took some of the first photomicrographs of metals.

source A source of *light* or other *radiation*.

source, extended A *source* whose dimensions are large enough to ensure that the emitted *radiation* has a very low *degree of coherence*.

source, point A *source* whose dimensions are sufficiently small to cause the emitted *radiation* to have a very high *degree of coherence*.

source, primary A surface, space, or component emitting *light* produced by a transformation of energy.

source, secondary (substitute) A surface, space, or component which is not self-emitting but receives *light*, e.g. in the form of a *real image* of a *primary source*, and redirects it, at least in part, by *reflection*, *transmission*, or *scattering*.

source-focused illumination *See* **illumination, source-focused**

space, Gaussian The hypothetical cylindrical narrow space closely surrounding the *optical axis*. It is synonymous with the expression *paraxial* region. (*See also* **ray, paraxial**)

space, image The space on that side of an optical system where the *image* is located. In the case of *reflection* or formation of a *virtual image* this space may coincide with the *object space*.

space, object The space on that side of an optical system where the *object* is located. In reflection or formation of a *virtual image* this space may coincide with the *image space*.

space, ray The total space between the *object* and the *image* through which *rays* pass from the *object* to the *image*.

spatial Relating to three-dimensional space.

spatial vision *See* **vision, spatial**

specific rotation *See* **rotation, specific**

specimen That which is to be studied with a *microscope*.

spectral Related to *wavelength* or being a function of wavelength.

Note: (1) Wavelength-dependent optical properties, such as refractive index, transmittance and reflectance are referred to as spectral when they are quoted for monochromatic radiation. They may be designated by the appropriate symbol for the quantity, with the subscript λ. (2) A short term for spectral concentration, indicated by the subscript λ. Care must be taken to distinguish between the quantities defined in (1) and (2); spectral concentration may also be wavelength dependent.

spectral concentration (or spectral density) [photom.] The rate of change of any *photometric* or *radiometric quantity* with the *wavelength*.

Note: The wavelength may be replaced by another quantity, specifying vibration, frequency, wave number, etc.

spectral luminosity curve *See* **luminosity curve**

spectrophotometer [photom.] A *photometer* for measuring *spectral* properties.

spectrophotometry [photom.] The use of or investigations with a *spectrophotometer*.

spectrum A display produced by the separation of the *monochromatic* components of a *complex radiation*.

spectrum, continuous A *spectrum* in which the *intensities* of the *monochromatic components* are equal or change so that all wavelengths within a wide band are represented and there are no abrupt intensity changes between adjacent components.

spectrum, line A *spectrum* in which there are marked *intensity* changes between adjacent *monochromatic components*, i.e. it consists of a series of discrete lines or narrow bands.

spectrum, secondary *Colour* fringes surrounding features in the *image* which are due to differences in the *focal length* of a *lens* system which remain after *correction* of *chromatic aberration* for a standard pair of *wavelengths*.

spectrum, tertiary The minimal residual *chromatic aberration* still remaining in *apochromats*. It is normally not recognizable by subjective observation of the *image* but is only apparent when methods of testing (e.g. the *Abbe test plate*) are used.

spectrum locus [photom.] A line on a *chromaticity diagram* connecting the x and y *chromaticity co-ordinates* of the *spectral* colours. (See Fig. 10.)

specular reflection *See* **reflection, specular**

spherical aberration *See* **aberration, spherical**

stage (microscope stage) The platform, at right angles to the *optical axis* of the *microscope*, which carries the *object*. It is often fitted with mechanical movements (as in a *mechanical stage*) to allow easy positioning of the object in the x and y axes and movement along, and rotation about, the z-axis. (See Fig. 1a.)

stage, centring A *rotating stage* fitted with provision for bringing its axis of rotation into coincidence with that of the *microscope*.

stage, cooling A *stage* fitted with means for reducing the temperature of the *object*.

stage, heating A *stage* fitted with means for raising the temperature of the *object*.

stage, integrating A *mechanical stage* equipped with a mechanism or an electric device used to facilitate *modal analysis* and the evaluation of stereological parameters.

stage, levelling A *stage* designed to hold a *polished section* so that its surface is normal to the *optical axis* of the *microscope*.

stage, mechanical A *stage* fitted with screw or rack mechanisms (sometimes provided with calibrations) to assist in precise translational movement of the *object* in the x and y directions. Used for systematic examination of the object by searching in a meandering path along the x–y co-ordinates. It may be manually or motor operated, attachable, or built into the *microscope stand*. In computer-controlled microscopes stepper motors are used to drive the stage.

stage, rotating A *stage* fitted with means for rotating the *object* with respect to the *optical axis* of the *microscope*. It may or may not be centrable and/or calibrated for measuring angles of rotation.

stage, scanning A *mechanical stage* electronically or electrically controlled to move the *object* in steps or continuously in a *raster* fashion.

stage, sliding A movable *stage* consisting of two flat plates, the upper of which can be moved smoothly in all direction in the x–y plane over the lower one, which is fixed to the *stand*. The ease of movement is regulated by the viscosity of the layer of grease which is used to connect the two plates.

stage, universal [pol.] A device mounted on to a *rotatable microscope stage* and equipped with a gimbal mechanism for movement of the *object*. These movements comprise calibrated tilting and rotation around three or four axes (according to the type of stage) in addition to the usual movements of the microscope stage. This stage enables the effect of *optical anisotropy* to be investigated at any direction of illumination or observation.

stage clip A flat spring used to hold a *slide* in contact with the *microscope stage*.

Note: The use of stage clips facilitates the precise movement of a slide with the fingers.

stage micrometer *See* **micrometer, stage**

stand (microscope stand) The chassis, composed of the base (which may contain an *illuminating system* for *transmitted light*), the *substage, stage*, and the *limb* on which the mechanical and optical parts of the *microscope* (e.g. the *body tube*, etc.) are carried. The limb may include the *tube* (or a part of it) and/or the *reflected-light illuminator*. (See Fig. 1.)

star test A procedure for evaluating the *spherical aberration correction* of *microscope objectives*, using a *point source* as the *object* (an artificial star).

steradian SI unit of measurement for a *solid angle* (symbol sr).

stereology [quant.] A group of mathematical methods relating three-dimensional parameters describing a structure to two-dimensional measurements obtained from sections of the structure.

stereomicroscope *See* **microscope, stereo-**

stereoscopic vision *See* **vision, stereo**

stereoscopy Technical methods for using *stereo vision* as an aid for perception and measurement.

* **stilb** An obsolescent unit of *luminance* expressed in *candelas* per cm^2.

stop A term (often used loosely) for a *diaphragm*, usually of fixed size.

stop, central A *diaphragm* with a central round opaque area concentric with the *optical axis*. Used, for example, in the *front focal plane* of the *condenser* or in the *back focal plane* of the *microscope objective* for the *darkground* observation.

strabismus [physiol.] Squinting; deviation of the position of the *eyes* with respect to one another in which the *visual axis* of only one eye is aligned with the fixed point of observation. It may be temporary or permanent.

straight extinction *See* **extinction, straight**

stray light *See* **light, stray**

sub-microscopic Of a size below the *resolution* capability of the *light microscope*.

substage The assembly of mechanical and opto-mechanical parts attached to the *stand* of a *transmitted-light microscope* before the *stage*. It comprises the *condenser* with its carrier, *filter tray*, and (optionally) a *polarizer* with its carrier and/or *relay lenses* with their carriers. (See Fig. 1.)

substage, Abbe A *substage* fitted with a mechanism for lateral displacement and rotation of the *illuminating aperture diaphragm*.

substage condenser *See* **condenser, substage**

subtraction position *See* **position, subtraction**

surrounding field *See* **field, surrounding**

symmetrical extinction *See* **extinction, symmetrical**

T

telescope, auxiliary A two-stage *magnifier*, designed for use in place of the *eyepiece* to enable an *image* of the *back focal plane* of the *objective* to be inspected. Used principally for adjustment of the *microscope* illuminating system, especially with *phase contrast*. May also be used for *conoscopic observation*.

television head *See* **head, television**

television microscope *See* **microscope, television**

temperature, colour [photom.] The temperature (symbol T_c) in kelvin of the surface of an ideal *black body* which emits *radiation* of the same *chromaticity* as that from the *source* being described.

Note: The relative spectral distributions of the radiation of the black body and the source under consideration may not be similar: if so the chromaticity is said to be metameric.

temperature, distribution [photom.] A special case of *colour temperature* in which the *spectral* distribution of the *radiation* of the *black body* and the *source* under consideration are similar, i.e. for which the ordinates of the spectral distribution of radiation of the black body are proportional (or approximately so) to those of the source of light under consideration.

tertiary spectrum *See* **spectrum, tertiary**

test object An *object* of known, often standard, dimensions, microscopical structure, and *contrast*, used to assess the performance of a *microscope* system.

thickness, optical *See* **pathlength, optical**

tissue, lens (lens paper) A soft paper, manufactured especially for cleaning the surfaces of optical components without causing them damage by abrasion.

top lens *See* **lens, top**

total internal reflection *See* **reflection, total internal**

total magnification *See* **magnification, total**

transillumination *See* **illumination, transmitted-light**

transfer lens A *relay lens*.

transmission [photom.] Passage of *radiation* through a *medium* without change in the proportions of the *monochromatic* components of which the radiation is composed.

transmittance [photom.] The measure of *transmission* in terms of either the ratio of transmitted *intensity* to incident intensity (total transmittance) or the ratio of the intensity reaching the final or exit surface of the medium to that entering the medium (internal transmittance). Consequently, total transmittance is influenced by both *absorption* and *reflection*; internal transmittance is influenced by absorption only.

transmittance, luminous *See* **luminous transmittance**

transmitted light *Light* which passes through an *object*.

transmitted-light microscope *See* **microscope, transmitted-light**

transmitted-light microscopy *See* **microscopy, transmitted-light**

travelling microscope *See* **microscope, measuring**

trinocular tube *See* **tube, trinocular**

tristimulus colour values *See* **colour values, tristimulus**

tube That part of the *microscope* which connects the *objective* and the *eyepiece* (see Fig. 1).

Note: In early microscopes the tube was in the form of a hollow cylinder carrying at one end the objective-locating surface and at the other the eyepiece-locating surface. In microscopes of more recent design the tube may be divided into two or more sections or housings, one or more being attached to the stand. The housings may not be cylindrical but shaped to offer the most convenient manipulation of the included opto-mechanical elements. In the case of a reflected-light microscope that part of the reflected-light illuminator which is situated between the nosepiece and the primary image plane is considered to be part of the tube. (*See also* **tube, body; tube, intermediate; tube, viewing; tube head** and Fig. 2.)

tube, binocular A *viewing tube* designed to accept two *eyepieces* for *binocular* viewing.

tube, body That part of the *tube*, fixed to or incorporated into the *stand,* containing the *nosepiece* on one side and carrying the *intermediate tube, viewing tube,* or *tube head* on the other. For certain purposes it may contain optical or opto-mechanical elements, e.g. *intermediate lenses, beam splitter, magnification changer, reflector* or *reflected-light illuminator, Bertrand lens*, mechanisms for operating *filters, retardation plates,* etc. For work with *infinity-corrected objectives*, it may contain the *tube-lens*. (See Fig. 1.)

* **tube, camera** *See* **head, camera**

tube, intermediate An optional part of the *tube*. It is a housing, either integral with the *stand* or exchangeable, forming part of the tube and containing some opto-mechanical elements (e.g. *magnification changer, filter tray, Bertrand lens, analyser*, slots for holding *retardation plates, beam splitter*, etc.). (See Fig. 1.)

tube, monocular A *viewing tube* designed to accept only one *eyepiece.*

* **tube, photometer** *See* **head, photometer**

tube, trinocular A *viewing tube* designed to accept two *eyepieces* for *binocular* viewing, together with a third eyepiece or other *lens* to enable simultaneous and/or alternate viewing and other use of the *image* (e.g. in *photomicrography*).

tube, viewing A section of the *tube* equipped to carry one or more *eyepieces.* For use with *infinity-corrected objectives* it may contain the *tube-lens*. It is limited at one end by the *eyepiece-locating surface* and at the other by the *body-tube-locating surface*. (See Figs. 1 and 2.)

tube factor The factor by which the *lateral magnification* of the *primary image* is changed by an *intermediate lens* inserted between the *objective* and the primary image or by a *tube-lens* whose focal length is different from the appropriate one.

tube head That part of the *tube* which joins the *body tube* or *intermediate tube* to any light-receiving device (e.g. photographic camera, television camera, photometric detector, viewing screen). (See Fig. 1.)

tube-lens An *intermediate lens* designed to operate as an essential component of *infinity-corrected objectives* and located in either the *body tube* or the *viewing tube*. The tube-lens should be considered to be part of the objective lens system and will thus influence the effective *magnifying power* and possibly the state of *correction* of the system.

Note: The tube-lens which is taken as the basis for the nominal magnification of an objective is called the 'tube-lens of appropriate focal length'.

tubelength, mechanical One of the *optical fitting dimensions*. For *objectives* corrected for a finite *primary image distance* it is the length of the *tube* in its simplest form (i.e. without any *intermediate lenses*) and is the distance in air between the *objective-locating surface (of the nosepiece)* and the *eyepiece-locating surface (of the viewing tube)*. It commonly has a value of 160 mm. For *infinity-corrected objectives* the mechanical tubelength is hypothetically considered to be infinite. (See Fig. 2 and Table 2.)

tubelength, optical The distance between the *back focal plane* of the *objective* and the *primary image plane* (which is the *front focal plane* of the *eyepiece*). This distance is *not* one of the *optical fitting dimensions* of the *microscope* and is relevant only to tubes fitted with objectives corrected for a finite primary image distance.

tubelength correction lens *See* **lens, tubelength correction**

twilight vision *See* **vision, twilight**

U

ultramicroscopy The detection of *submicroscopic* particles by means of *darkground microscopy* using a very intense *light*.

ultraviolet microscopy *See* **microscopy, ultraviolet**

ultraviolet radiation *See* **radiation, ultraviolet**

uncrossed polars *See* **polars, uncrossed**

undercorrection An error in *correction* of *spherical aberration*, leading to an unsharp *image* with some lack of *contrast*. In *microscopy* it may be caused by the use of a *cover glass* thinner than, or a *mechanical tubelength* shorter than, the values assumed in the computation of the *objectives*.

uniaxial, optically [pol.] Having one *crystal optical axis*.

unilateral oblique illumination *See* **illumination, unilateral oblique**

universal stage *See* **stage, universal**

useful magnification *See* **magnification, useful range of**

V

vector, light A short term for the electric vector of *light*. It is the parameter which causes the sensation of *brightness*.

velocity, phase *See* **phase velocity**

vergence [physiol.] Simultaneous movement of both *eyes* in opposite directions so as to alter the angle between their *fixation lines*.

version (of the eyes) [physiol.] Movement of the two *eyes* in the same direction, as when tracking a moving *object*.

vertex The intersection of the *optical axis* and the surface of a *lens*.

vertical illumination *See* **illumination, vertical**

vibration direction *See* **direction, vibration**

vibration directions, principal *See* **directions, vibration, principal**

vibration plane *See* **plane, vibration (of electromagnetic radiation)**

video-enhanced contrast microscope *See* **microscope, video-enhanced contrast**

video-enhanced contrast microscopy *See* **microscopy, video-enhanced contrast**

viewing angle The angle subtended by an *object* or a *field* at the *eye*. (See Fig. 7b.)

viewing tube *See* **tube, viewing**

virtual image *See* **image, virtual**

visible radiation *See* **light**

vision, daylight (or photopic) [physiol.] Vision adapted to *luminances* at which the *eye* has reached the photopic *spectral* response. For the normal eye it begins when the luminance is higher than about 10 cd m^{-2}.

vision, mesopic *See* **vision, twilight**

vision, night (or scotopic) [physiol.] Vision adapted to *luminances* at which the *eye* has reached the scotopic *spectral* response, i.e. when the luminance is less than about 10 cd m^{-2}.

vision, photopic *See* **vision, daylight**

vision, scotopic *See* **vision, night**

vision, spatial [physiol.] The visual perception of depth: the ability of the *visual organ* to recognize different distances of *object* points from the observer. It consists of *stereo vision* and *non-disparate spatial vision*.

vision, spatial, non-disparate [physiol.] Spatial localization from a *monocular image*; the perception of depth resulting from either monocular or *binocular* vision. Such spatial localization may be due to geometrical perspective, distribution of light and shadow, object overlap (overlap perspective), atmospheric conditions (air perspective), movement parallax, or relation between the size of the retinal image and the size of the object which is imaged.

vision, stereo (stereoscopic) [physiol.] The perception of depth due to the formation on the two *retinae* of disparate *images* of *object* points. It is only possible with *binocular* vision.

vision, twilight (or mesopic) [physiol.] Vision at the boundary between *daylight vision* and *night vision*.

visual acuity *See* **acuity, visual**

visual angle [physiol.] The *viewing angle*.

visual axis (of the eye) [physiol.] The line (*fixation line*) connecting a *fixation point* in the *object* with its *image* on the *retina*. This line passes approximately through both *nodal points* of the *eye*.

visual field *See* **field of view**

visual field diaphragm *See* **diaphragm, visual field**

visual organ *See* **organ, visual**

vitreous body (of the eye) [physiol.] The transparent jelly-like substance filling the posterior part of the eyeball, behind the *lens* (of the *eye*).

W

warm chamber *See* **chamber, warm**

wave number The reciprocal of the *wavelength*($^1/_\lambda$ m^{-1} or cm^{-1}), where λ is expressed in metres or centimetres.

wave optics *See* **optics, wave**

wavegroup (wavepacket) Several waves superimposed, having their *wavelengths* distributed within a narrow range and travelling in the same direction.

wavelength The distance on a periodic wave between two successive points at which the *phase* is the same. Represented by the symbol λ, and usually expressed in nanometres.

wavelength, central [photom.] The *wavelength* of the central component of a *wavelength band*.

Note: If the profile of the band has a symmetrical shape, the central wavelength is the same as the peak wavelength.

wavelength, complementary dominant [photom.] That *wavelength* indicated by the intersection of the *spectrum locus* with the extension of a straight line drawn from the *chromaticity co-ordinates* of a standard *source* to those of the sample if the original line intersected the *purple line* on the *chromaticity diagram* rather than the spectrum locus itself. (See Fig. 10.)

wavelength, dominant [photom.] A measure of *hue*. It is given by the *wavelength* corresponding to the intersection of the *spectrum locus* and the extension of a straight line from the *chromaticity coordinates* of a standard *source* to those of a sample. (See Fig. 10.)

wavelength, peak [photom.] The *wavelength* in a *wavelength band* which corresponds to the *spectral* position of maximum *intensity*.

wavelength band The width of the band of *wavelengths* lying between two specified wavelengths.

wavetrain A sequence of finite length consisting of *light waves* of equal *phase, amplitude, wavelength,* and *polarization state,* travelling in the same direction with equal velocity.

wedge, quartz [pol.] A *retardation plate* consisting of a wedge of quartz (or two such wedges in the *subtraction position*) producing *optical pathlength differences* continuously variable between 0 and 3 or 4λ (with λ being taken as about 550 nm) along the length. This property results in the production of a series of *interference colours* in the form of fringes perpendicular to the length of the wedge.

Note: With monochromatic light the coloured fringes are seen as alternating dark and bright bands.

Wenham, Francis H. (1824–1908) An English marine engineer and enthusiastic amateur microscopist who invented many accessories for the microscope including a binocular tube which bears his name.

white body (or white standard) [photom.] A *sample* showing *diffuse reflection* and having a *reflectance* of 1 (or 100%) or a perfectly transparent sample with internal *transmittance* of 1 (or 100%) over the whole visible *spectrum*.

white light *Light* containing any mixture of *monochromatic radiation* in such proportions that it will be perceived as without *chromatic colour*.

white standard *See* **white body**

wide field eyepiece *See* **eyepiece, wide field**

width, half-height *See* **half-peak-height bandwidth**

Wollaston, William Hyde (1766–1828) An English scientist who carried out research in physics, crystallography, and astronomy. The inventor of a goniometer and the polarizing prism which bears his name.

Wollaston prism *See* **prism, Wollaston**

working distance *See* **objective, free working distance of**

Z

Zernike, Frits (1888–1966) A Dutch Professor of theoretical physics at the University of Groningen. In 1933 he published his invention of phase-contrast microscopy, for which he was honoured with the Nobel prize for physics in 1953.

zone lens *See* **plate, zone**

zone plate *See* **plate, zone**

zoom *See* **pancratic**

Appendix I

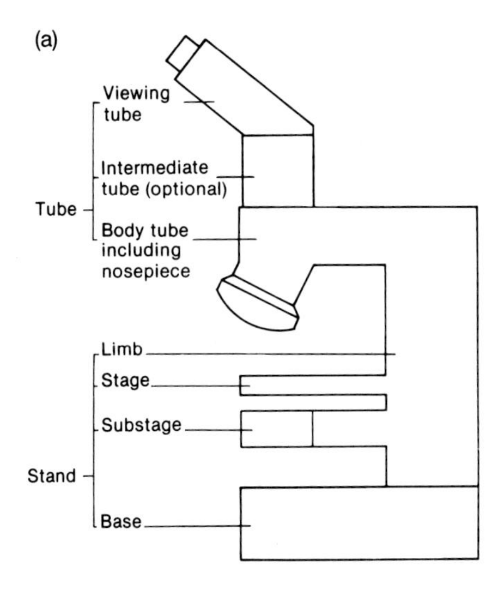

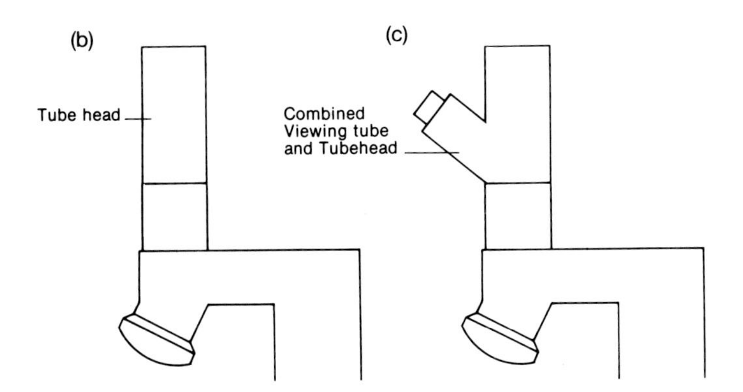

Fig. 1. (a) The main modules of the microscope. (b) Alternative with tube head in place of viewing tube. (c) Alternative with combined viewing tube and tubehead.

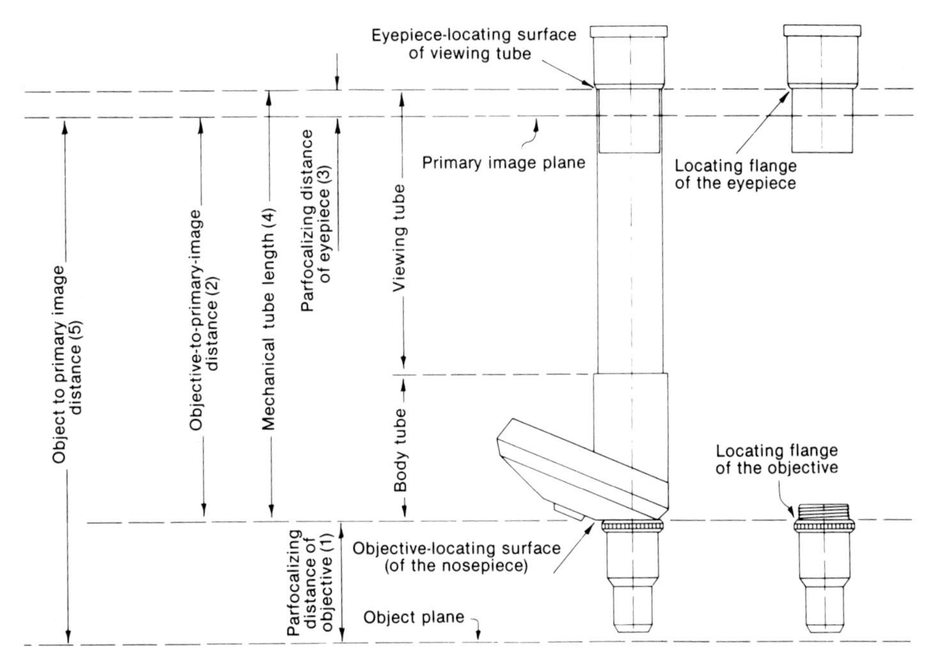

Fig. 2. Locating surfaces, reference planes, and optical fitting dimensions.
Standardized values of distances (see also Table 2):

In microscopes with objectives corrected for a finite objective to primary image distance (1) = 45 mm, (2) = 150 mm, (3) = 10mm, (4) = 160 mm, (5) = 195 mm.

In microscopes with infinity-corrected objectives (4) is infinite (hypothetical), (2) is infinite (hypothetical), (3) = 10 mm, (1) = 45 mm, (5) is infinite (hypothetical).

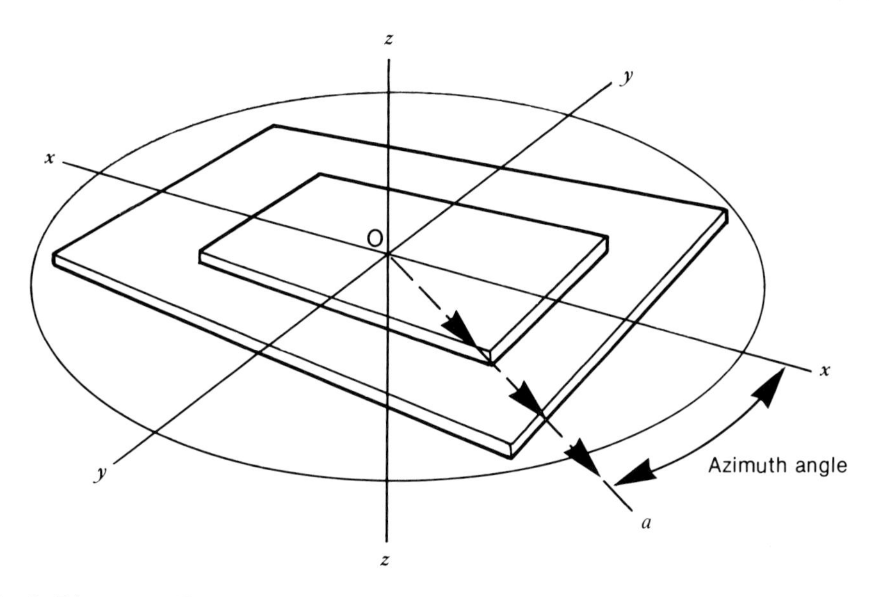

Fig. 3. Diagram to illustrate the relationship between azimuth angle and reference directions. Directions x and y are those in which the stage may be traversed, and z is the direction in which focusing adjustments are made. Angle aOx is the azimuth angle which describes direction Oa.

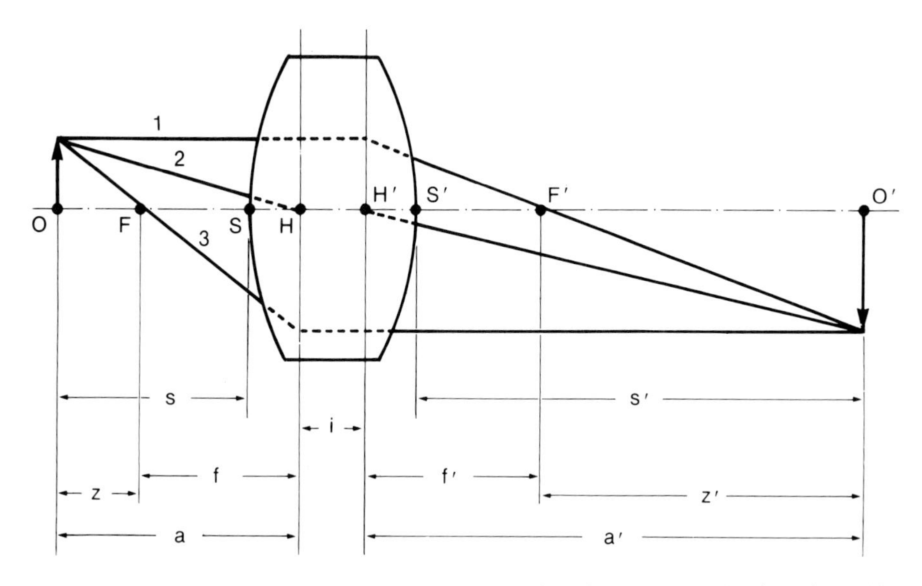

Fig. 4. Cardinal elements of a lens in air. O, object point (plane); O′, image point (plane); F, object-side focal point (plane); F′, image-side focal point (plane); H, object-side principal point (plane); H′, image-side principal point (plane); S, object-side lens vertex; S′, image-side lens vertex; f, object-side focal length; f', image-side focal length; z, object-side focus to object distance; z', image-side focus to image distance; a, object distance; a', image distance; i, hiatus.

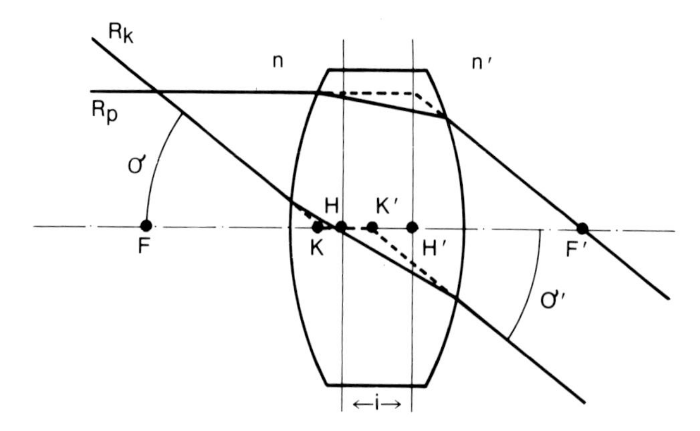

Fig. 5. Cardinal elements of a lens bounded by media of different refractive indices (n and n'). F, object-side focal point (plane); F′, image-side focal point (plane); H, object-side principal point (plane); H′, image-side principal point (plane); K, object-side nodal point (plane); K′, image-side nodal point (plane); R_p, ray parallel to axis; R_k, nodal point ray; σ, entrance angle of nodal point ray; σ', exit angle of nodal point ray; n, refractive index of medium bounding lens on object side; n', refractive index of medium bounding lens on image side; i, hiatus.

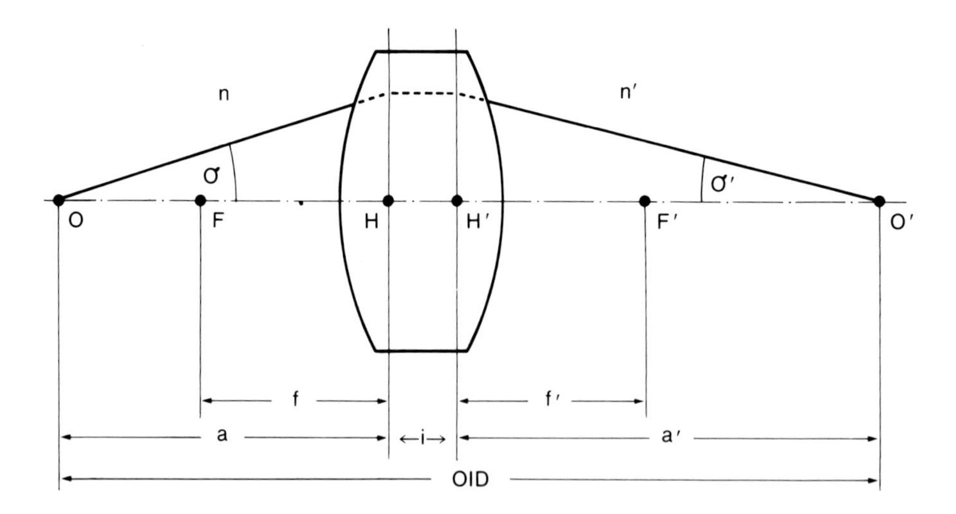

Fig. 6. The sine condition:

$$\frac{n \sin \sigma}{n' \sin \sigma'} = M_l = \text{constant}$$

M_l, lateral magnification; O, object point; O′, image point; F, object-side focal point; F′, image-side focal point; H, object-side principal point; H′, image-side principal point; f, object-side focal length; f', image-side focal length; a, object distance; a', image distance; OID, object to image distance; i hiatus (distance between H and H′); σ, object-side angle between a marginal ray and the axis; σ', image-side angle between this ray and the axis; n and n', refractive indices of media adjacent to lens.

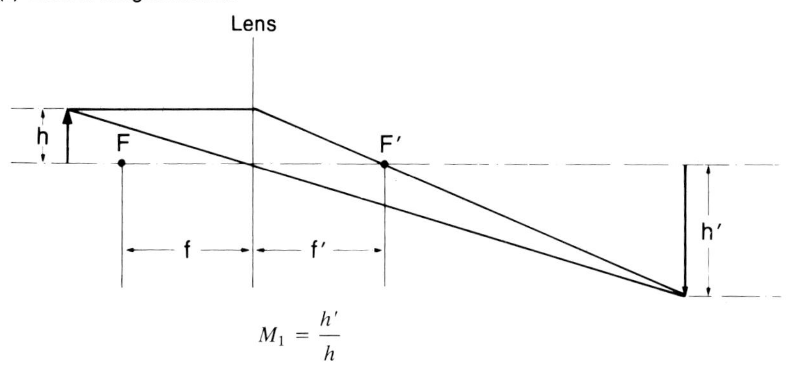

$$M_1 = \frac{h'}{h}$$

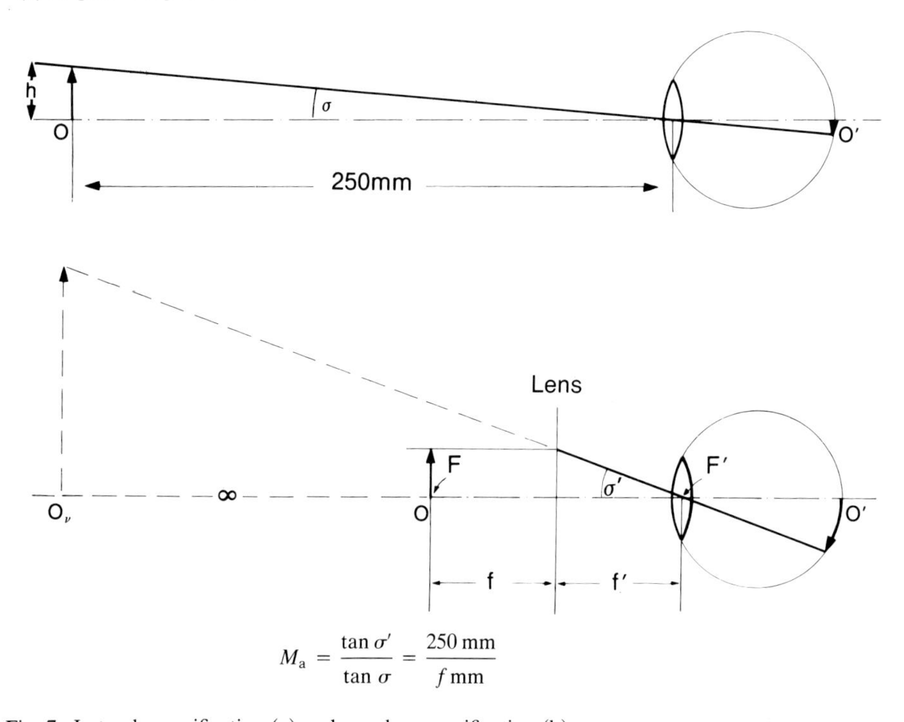

$$M_a = \frac{\tan \sigma'}{\tan \sigma} = \frac{250 \text{ mm}}{f \text{ mm}}$$

Fig. 7. Lateral magnification (a) and angular magnification (b).

M_a, angular magnification; M_1, lateral magnification; h, object size; h', image size; σ, viewing angle subtended by distance h at the reference viewing distance from the eye; σ', viewing angle which the virtual image of h, formed by the magnifier, appears to subtend; f, object-side focal length; f', image-side focal length (the values of the focal lengths on the image side and the object side are taken to be the same); 250 mm, reference viewing distance.

In this diagram for simplicity the lens is considered to be infinitely thin and thus represented as a line; both its principal planes coincide with this line.

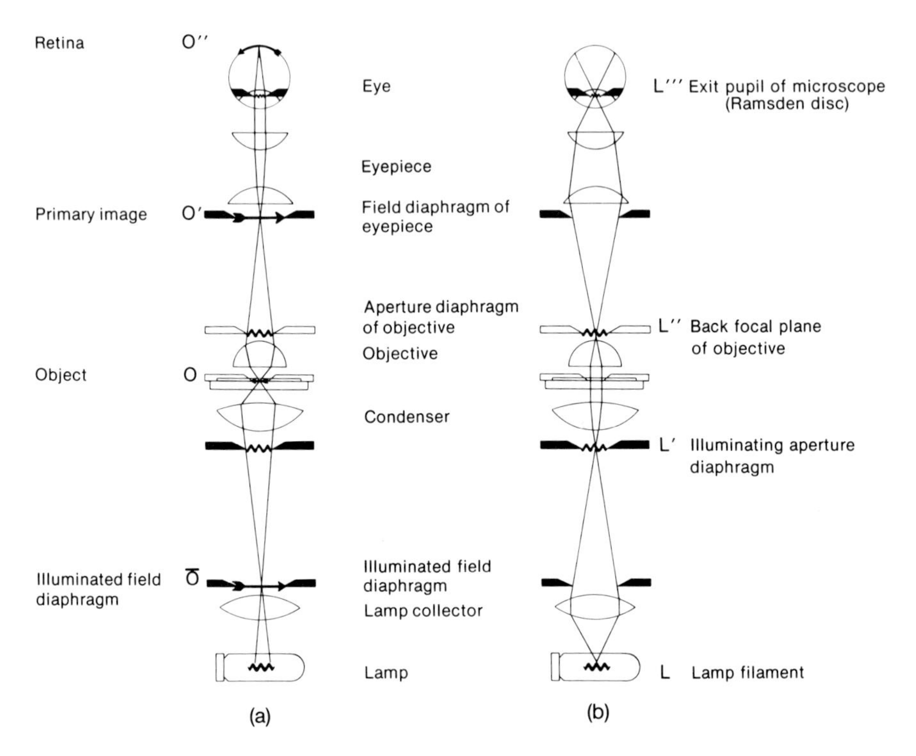

Fig. 8. Ray paths in a transmitted light microscope.

Two sets of conjugate planes are shown. Set O in Fig. 8a is conjugate with the object (O) and with field diaphragm planes (fields), and set L in Fig. 8b is conjugate with the lamp filament (L) and with aperture diaphragm planes (pupils). Subsequent images of object and filament are shown as O′, O′′, L′, L′′, L′′′, L′′′′. Ō is the 'back image' of the object on the illuminated field diaphragm.

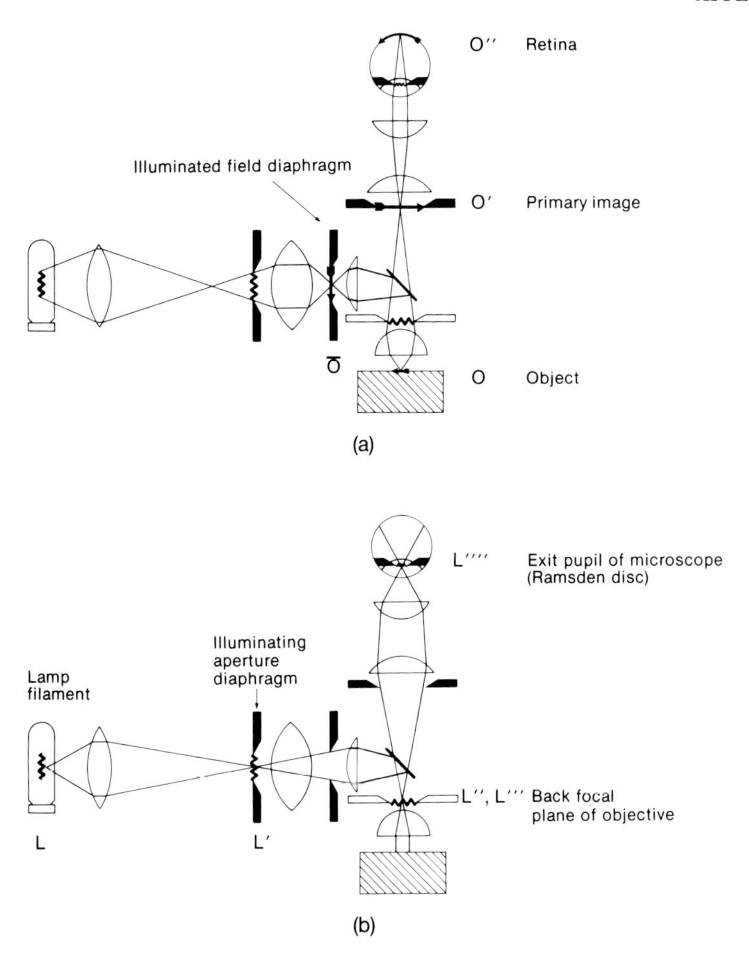

Fig. 9. Ray paths in a reflected light microscope.

Two sets of conjugate planes are shown. Set O in Fig. 9a is conjugate with the object (O) and with field diaphragm planes (fields), and set L in Fig. 9b is conjugate with the lamp filament (L) and with aperture diaphragm planes (pupils). Subsequent images of object and filament are shown as O′, O′′, L′, L′′, L′′′L′′′′. Ō is the 'back image' of the object on the illuminated field diaphragm.

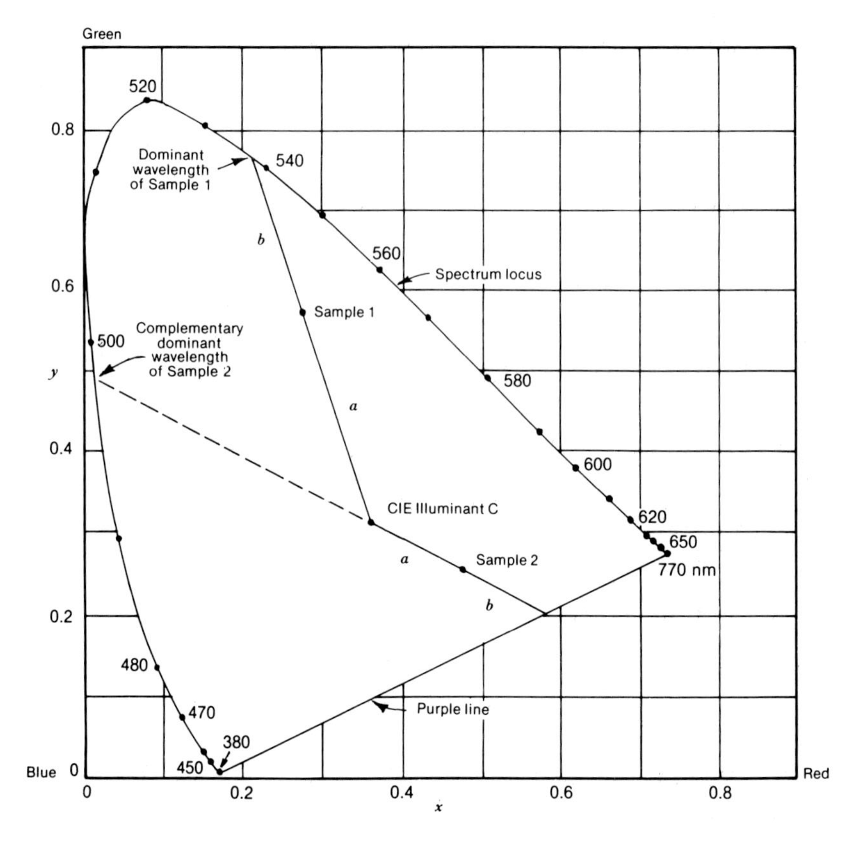

Fig. 10. Chromaticity diagram, showing the horseshoe-shaped spectrum locus with the purple line joining its ends, and the positions of the x and y chromaticity co-ordinates for two samples, 1 and 2, and for a standard light source (e.g. CIE Illuminant C, which simulates overcast sky daylight). A line may be drawn from the co-ordinates of the standard source to those of the sample. If this line intersects the spectrum locus (as with sample 1), this intersection is known as the dominant wavelength; if, however, it intersects the purple line (as with sample 2), its extension to intersect the spectrum locus indicates the complementary dominant wavelength.

TABLE 1. Radiometric and Corresponding Photometric Quantities

Radiometric quantities			Photometric quantities		
Quantity	Symbol	SI unit	Quantity	Symbol	SI unit
Radiant energy	$Q_e = \int \Phi_e \, dt$	J = W s	Quantity of light	$Q_v = \int \Phi_v \, dt$	lm s
Radiant flux	$\Phi_e = \frac{d\,Q_e}{dt}$	W	Luminous flux	$\Phi_v = \frac{d\,Q_v}{dt}$	lm
Radiant intensity	$I_e = \frac{d\,\Phi_e}{d\,\Omega}$	$W\,sr^{-1}$	Luminous intensity	$I_v = \frac{d\,\Phi_v}{d\,\Omega}$	$cd = lm\,sr^{-1}$
Radiance	$L_e = \frac{d^2\,\Phi_e}{d\,\Omega\,dA\,\cos\theta}$	$W\,sr^{-1}\,m^{-2}$	Luminance	$I_v = \frac{d^2\,\Phi_v}{d\,\Omega\,dA\,\cos\theta}$	$cd\,m^{-2} = lm\,sr^{-1}\,m^{-2}$
Radiant flux (surface density	$\frac{d\,\Phi_e}{dA}$	$W\,m^{-2}$	Luminous flux (surface) density	$\frac{d\,\Phi_v}{dA}$	$lm\,m^{-2}$
Irradiance	$E_e = \frac{d\,\Phi_e}{dA}$	$W\,m^{-2}$	Illuminance	$E_v = \frac{d\,\Phi_v}{dA}$	$lm\,m^{-2} = lx$
Radiant exitance	$M_e = \frac{d\,\Phi_e}{dA}$	$W\,m^{-2}$	Luminous exitance	$M_v = \frac{d\,\Phi_v}{dA}$	$lm\,m^{-2}$
Radiant exposure	$H_e = \frac{dQ_e}{dA} = \int E_e \, dt$	$J\,m^{-2}$	Light exposure	$H_v = \frac{dQ_v}{dA} = E_v \, dt$	$lm\,s\,m^{-2}$

Other symbols used above: A, area; θ, directional angle; Ω, solid angle; t, time (duration).

SI units: J, joule; W, watt; s, second; m, metre; sr, steradian; cd, candela; lm, lumen; lx, lux.

The subscripts v (related to vision) or e (related to energy) are used if required, i.e. unless the character of the quantity in consideration is obvious from the context.

TABLE 2. Standardized and other dimensions important for the proper use of the microscope. See also Fig. 2.

(a) Objective screw thread dimensions (in mm).

Dimensions		Ext. diam.	Pitch diam.	Core diam.	Calc. play male/female	Allowance	Tolerance
Female	Max.	20.396	19.944	19.492		+0.076	
	Min.	20.320	19.868	19.416	Min. 0.046	0	0.076
Male	Max.	20.274	19.822	19.370	Max. 0.198	−0.046	
	Min.	20.198	19.746	19.294		−0.122	0.076

Based on data given in the specification by the RMS (*J. R. Microsc. Soc.* **56**, 377–80 (1936)) and now standardized in ISO Standard 8038 (1985) and in BS 7012, pt. 4.

(b) Dimensions for microscopes equipped with objectives corrected for a finite primary image distance or an infinite image distance.

	For finite image distance	For infinite image distance
†Mechanical tubelength	160mm	Optically infinite (∞). The actual length of the tube is fixed by the microscope designer, taking into account the focal length of the tube-lens, the distance of the tube-lens from the objective locating surface, and the eyepiece parfocalizing distance
†Objective to primary image distance	150 mm	Optically infinite (∞). The real distance is the difference between the actual length of the tube and the eyepiece parfocalizing distance
Object to primary image distance	195 mm	Optically infinite (∞). The real distance is the sum of the actual objective to primary image distance and the objective parfocalizing distance

†This symbol indicates a standardized dimension (see BS 7012, pt. 3).

(c) The following dimensions are the same for microscopes fitted with either finite or infinity-corrected objectives.

[1] Objective parfocalizing distance	45 mm (uncovered specimen); 45.06 (specimen with 0.17 mm cover glass)
†Eyepiece parfocalizing distance	10 mm
†Eyepiece outside diameter	23.2 mm or 30 mm
Inside diameter of tube	23.2 mm or 30 mm
Field-of-view no. of eyepiece	Depends on individual eyepiece (marked in mm on its top)
Objective, free working distance	Depends on individual objective
Condenser, free working distance	Depends on individual condenser
[2] Cover glass, thickness of	No. 1 (general purpose) 0.17 mm ± 0.04mm No. 1H (high performance) 0.17 mm ± 0.02 mm
[3] Microscope slide, thickness of	Three standard thicknesses: 0.9 ± 0.1 mm 1.0 ± 0.1 mm 1.1 ± 0.1 mm
[4] Immersion oil	Refractive index n_e 1.518 ± 0.0005 Abbe number = 44 ± 3
[5] Light filters	**Form A** Filters to be used in either the path of image rays or for the illuminating beam Diameters: 18 ± 0.11 mm, 25 ± 0.13 mm **Form B** Filters to be used solely in the illuminating beam Diameters: 18 ± 0.11 mm, 25 ± 0.13 mm, 32 ± 0.16 mm, 50 ± 0.16 mm Parallelism, surface form error, surface quality, etc. tolerances vary between Form A and Form B filters
[6] Graticules for eyepieces	Diameter (mm) 19, 21, 26 ± 0.033 mm Thickness 1.5 mm + 0.2, − 0.2 mm Protective chamfer 0.1 mm + 0.1 mm

†This symbol indicates a standardized dimension (BS 7012, pt. 3).

1 BS 7012, pt. 3   4 BS 7011, pt. 1
2 BS 7011, pt. 3   5 BS 7012, pt. 6
3 BS 7011, pt. 2   6 BS 7012, pt. 8

Note: The ISO standards are not yet finalized.

Appendix II

Equivalent terms: English–French–German

English	French	German
ABBE CONDENSER	Condenseur d'Abbe	Abbescher Kondensor
ABBE DIFFRACTION APPARATUS	Appareil de diffraction d'Abbe	Abbescher Diffraktionsapparat
ABBE NUMBER	Nombre d'Abbe; Constringence	Abbesche Zahl
ABBE SUBSTAGE	Appareil d'éclairage d'Abbe	Abbescher Beleuchtungsapparat
ABBE TEST PLATE	Lame-test d'Abbe	Abbesche Testplatte
ABBE THEORY (OF IMAGE FORMATION)	Théorie de la formation des images d'après Abbe	Abbesche Theorie der Bildentstehung
ABERRATION	Aberration	Aberration; Abbildungsfehler
ABERRATION, APERTURE	Aberration sphérique	Öffnungsfehler
ABERRATION, CHROMATIC	Chromatisme	chromatische Aberration
ABERRATION, CHROMATIC AXIAL	Chromatisme longitudinal; (axial)	axiale chromatische Aberration
ABERRATION, CHROMATIC LATERAL	Chromatisme de grandeur; (latéral)	laterale chromatische Aberration
ABERRATION, SPHERICAL	Aberration sphérique	sphärische Aberration
ABERRATIONS, MONOCHROMATIC	Aberrations géométriques	monochromatische Abbildungsfehler
ABSORBANCE	Densité optique	Extinktion
ABSORPTION	Absorption	Absorption
ACCOMMODATION	Accommodation	Akkommodation
ACCOMMODATION DISTANCE	Distance d'accommodation	Akkommodationsentfernung
ACCOMMODATION RANGE	Amplitude d'accommodation	Akkommodationsbereich
ACCOMMODATION STATE	État d'accomodation	Akkommodationszustand
ACHROMAT	Achromat	Achromat
ACHROMATIC–APLANATIC CONDENSER	Condenseur achromatique et aplanétique	aplanatisch–achromatischer Kondensor
ACHROMATIC COLOUR	Couleur achromatique	unbunte Farbe
ACTIVITY, OPTICAL	Activité optique; Pouvoir rotatoire	optische Aktivität; optisches Drehvermögen
ACUITY, VISUAL	Acuité visuelle	Sehschärfe
ADAPTATION	Adaptation	Adaptation
ADAPTATION STATE	État d'adaptation	Adaptationszustand
ADDITION POSITION	Position avec axes parallèles	Additionsstellung
ADJUSTMENT, COARSE	Mouvement rapide	Grobtrieb
ADJUSTMENT, FINE	Mouvement lent	Feintrieb
ADJUSTMENT LENGTH	Longueur d'équilibrage	Abgleichlänge
AERIAL IMAGE	Image aérienne	Luftbild
AFTER IMAGE	Impression rémanente	Nachbild
AIRY DISC	Tache d'Airy	(Airysches) Beugungsscheibchen
AKEHURST SLIDE		
AMETROPIA	Amétropie	Ametropie; Fehlsichtigkeit
AMPLIFIER	Oculaire divergent de projection	
AMPLITUDE	Amplitude	Amplitude
ANALYSER	Analyseur	Analysator

ANALYSIS, MODAL		Modalanalyse
ANGLE, AZIMUTH	Azimuth	Azimutwinkel
ANGLE, CONVERGENCE	Convergence	Konvergenzwinkel
ANGLE, CRITICAL	Angle critique	kritischer Winkel
ANGLE, DIVERGENCE	Divergence	Divergenzwinkel
ANGLE, EXTINCTION	Angle d'extinction	Auslöschungswinkel
ANGLE, OPTIC AXIAL	Angle des axes optiques dans un cristal biaxe	optischer Achsenwinkel
ANGLE, PHASE	Phase	Phasenwinkel
ANGLE, SOLID	Angle solide	Raumwinkel
ANGLE, VIEWING	Angle visuel; Diamètre apparent	Blickwinkel; Betrachtungs winkel
ANGLE, VISUAL	Angle visuel; Diamètre apparent	Gesichtswinkel; Sehwinkel
ANGLE OF INCIDENCE	Angle d'incidence	Einfallswinkel
ANGLE OF REFLECTION	Angle de réflexion	Reflexionswinkel
ANGLE OF VIEW (OF EYEPIECE)	Champ angulaire (d'un oculaire)	Bildwinkel (des Okulars)
ÅNGSTRÖM UNIT	Ångström	Ångström-Einheit
ANGULAR APERTURE	Ouverture angulaire	Öffnungswinkel
ANGULAR MAGNIFICATION	Grossissement commercial	Angularvergrößerung; Winkelvergrößerung
ANISOTROPISM	Anisotropie	optische Anisotropie
ANISOTROPY	Anisotropie	Anisotropie
ANISOTROPY, OPTIICAL	Anisotropie optique	optische Anisotropie
ANNULAR ILLUMINATION	Eclairage annulaire	ringförmige Beleuchtung
APERTOMETER	Apertomètre	Apertometer
APERTURE	Ouverture	Apertur
APERTURE ABERRATION	Aberration sphérique	Öffnungsfehler
APERTURE, ANGULAR	Ouverture angulaire	Öffnungswinkel
APERTURE, IMAGE-SIDE (OF A MICROSCOPE OBJECTIVE)	Ouverture numérique dans l'espace image (d'un objectif de microscope)	bildseitige numerische Apertur (eines Mikroskop-Objektivs)
APERTURE, NUMERCAL	Ouverture numérique	numerische Apertur
APERTURE, OBJECT-SIDE (OF A MICROSCOPE OBJECTIVE)	Ouverture numérique (d'un objectif de microscope)	objektseitige numerische Apertur (eines Mikroskop-Objektivs)
APERTURE DIAPHRAGM	Diaphragme d'ouverture	Aperturblende
APERTURE PLANE	Plan pupillaire; Pupille	Aperturebene
APLANATIC	Aplanétique	aplanatisch
APOCHROMAT	Apochromat	Apochromat
APODIZATION	Apodisation	Apodisation
AQUEOUS HUMOR	Humeur aqueuse	Kammerwasser
AREAL	Surfacique	flächig; Flächen- (als Präfix)
AREAL MAGNIFICATION		Flächenvergrößerung
ARTEFACT	Artefact	Artefakt; Kunstprodukt
ASPHERICAL	Asphérique	asphärisch
ASPHERICAL LENS	Lentille asphérique	asphärische Linse
ASTIGMATISM	Astigmatisme	Astigmatismus
AUTOFLUORESCENCE	Autofluorescence	Eigenfluoreszenz
AUTORADIOGRAPHY	Autoradiographie	Autoradiographie
AUXILIARY MICROSCOPE	Viseur de Bertrand	Hilfsmikroskop
AUXILIARY TELESCOPE	Viseur de Bertrand	Einstellfernrohr; Hilfs-mikroskop
AXIAL CHROMATIC ABERRATION	Chromatisme longitudinal; Chromatisme de position	axiale chromatische Aberration
AXIAL FIGURE	Figure d'axes	Achsenbild
AXIAL ILLUMINATION	Éclairage axial	axiale Beleuchtung
AXIAL MAGNIFICATION	Grandissement axial	Axialvergrößerung
AXIAL RAY		Axialstrahl
AXIS, OPTIC	Axe optique (de l'oeil)	Sehachse
AXIS, OPTICAL	Axe optique	optische Achse
AXIS, CRYSTAL OPTICAL	Axe optique d'un cristal	kristalloptische Achse
AXIS, VISUAL (OF THE EYE)	Axe visuel (de l'oeil)	optische Achse (des Auges)
AZIMUTH ANGLE	Azimuth	Azimutwinkel

APPENDIX II

BACK FOCAL PLANE	Plan focal image	hintere Brennebene
BALSAM, CANADA	Baume du Canada	Kanadabalsam
BANDWITH, HALF-PEAK HEIGHT	Largeur de bande à mi-hauteur	Halbwertsbreite
BARREL DISTORTION	Distortion en barillet	tonnenförmige Verzeichnung
BARRIER FILTER	Filtre de blocage; Filtre d'arrêt	Sperrfilter
BASE, MICROSCOPE	Pied du microscope	Mikroskopfuß
BEAM	Faisceau	Lichtbündel
BEAM, INCIDENT	Faisceau incident	einfallendes Lichtbündel
BEAM, LIGHT	Faisceau de lumière	Lichtbündel
BEAM SPLITTER	Séparateur de faisceaux	Strahlungsteiler
BECKE LINE	Ligne de Becke	Beckesche Linie
BEER–LAMBERT LAW	Loi de Beer–Lambert	Lambert–Beersches Gesetz
BERTRAND DIAPHRAGM	Diaphragme de Bertrand	Bertrand-Blende
BERTRAND LENS	Lentille de Bertrand	Bertrand-Linse
BIAXIAL, OPTICALLY	Biaxe	optisch zweiachsig
BINOCULAR	Binoculaire	binokular
BINOCULAR HEAD	Tube binoculaire	Binokulartubus
BINOCULAR MICROSCOPE	Microscope binoculaire; Loupe binoculaire	Binokular-Mikroskop
BINOCULAR TUBE	Tube binoculaire	Binokulartubus
BIREFLECTANCE	Biréflectance	Doppelreflexionsgrad
BIREFLECTION	Effet de biréflectance	Doppelreflexion
BIREFRACTION	Double réfraction	Doppelbrechung
BIREFRINGENCE	Biréfringence	Betrag der Doppelbrechung
BIREFRINGENCE, SIGN OF	Signe de la biréfringence	Vorzeichen der Doppelbrechung
BISECTRIX	Bissectrice	Bisektrix
BLACK BODY	Corps noir	schwarzer Körper; Hohlraum-strahler
BLACK-BODY RADIATION	Rayonnement du corps noir	schwarze Strahlung; Hohlraumstrahlung
BLACK-BODY RADIATOR	Corps noir; Radiateur intégral	Planckscher Strahler; schwarzer Strahler; Hohlraum-Strahler
BLOOMING	Traitement de surface	Beschichtung
BODY TUBE	Corps	Mikroskoptubus; Tubuskörper
BODY-TUBE-LOCATING SURFACES	Faces d'appui sur le corps	Anlageflächen am Tubuskörper
BRIGHT-FIELD	Fond clair	Hellfeld
BRIGHT-FIELD ILLUMINATION	Éclairage en fond clair	Hellfeldbeleuchtung
BRIGHT FIELD MICROSCOPY	Microscopie en fond clair	Hellfeldmikroskopie
BRIGHTNESS	Luminosité	Helligkeit
BROAD-BAND-PASS FILTER	Filtre à large bande	Breitbandpaßfilter
BROWNIAN MOVEMENT	Mouvement Brownien	Brownsche Molekularbewegung
BULB	Ampoule	Lampenkolben; Lampe
BUNDLE, RAY	Faisceau de rayons	Strahlenbündel
BUNDLE, RAY, CONVERGING	Faisceau de rayons convergents	konvergentes Strahlenbündel
BUNDLE, RAY, DIVERGING	Faisceau de rayons divergents	divergentes Strahlenbündel
BUNDLE, RAY, PARALLEL	Faisceau de rayons parallèles	paralleles Strahlenbündel
CAMERA HEAD	Tête photographique	Photokopf
CAMERA LUCIDA	Chambre clair	Camera lucida;Zeichenprisma
CAMERA TUBE	Tube photographique	Phototubus
CANDELA	Candela	Candela
CARDINAL DISTANCES		Kardinalstrecken
CARDINAL ELEMENTS	Eléments cardinaux	Kardinalelemente
CARDINAL PLANES		Kardinalebenen
CARDINAL POINTS		Kardinalpunkte
CARDIOID CONDENSER	Condenseur cardioïde	Kardioidkondensor
CATOPTRIC	Catoptrique	katoptrisch
CATADIOPTRIC	Catadioptrique	katadioptrisch

CEMENT	Colle	Kitt (Feinkitt)
CENTRAL STOP	Obturation centrale	Zentralblende
CENTRAL WAVELENGTH	Longeur d'onde centrale	Zentralwellenlänge
CENTRING NOSEPIECE	Porte-objectif centrable	zentrierbarer Objektiv-wechsler
CHAMBER, COLD	Chambre froide	Kühlkammer
CHAMBER, COUNTING	Cellule de comptage	Zählkammer
CHAMBER, CULTURE	Chambre de culture	Kulturkammer
CHAMBER, WARM	Chambre chaude	Heizkammer
CHROMATIC	Chromatique	farbig; Farb- (als Präfix)
CHROMATIC ABERRATION	Chromatisme	chromatische Aberration
CHROMATIC COLOUR	Couleur chromatique perçue	bunte Farbe
CHROMATIC DIFFERENCE OF MAGNIFICATION	Chromatisme latéral; Chromatisme de grandeur	chromatische Vergrößerungs-differenz
CHROMATICITY	Chromaticité	Farbart
CHROMATICITY CO-ORDINATES	Coordonnées tri-chromatiques	Normfarbwertanteile; Norm-farbwertkoordinaten
CHROMATICITY DIAGRAM	Triangle des couleurs; Diagramme de chromaticité	Normfarbtafel; Farbdreieck
CIRCLE OF LEAST CONFUSION	Cercle de moindre diffusion	Zerstreungskreis
CIRCULARLY-POLARIZED LIGHT	Lumière polarisée circulaire	zirkular polarisiertes Licht
COARSE ADJUSTMENT	Mouvement rapide	Grobtrieb
COATING OF LENS SURFACE	Traitement de surface des lentilles	Beschichtung von Linsen-oberflächen
CODE, COLOUR, OF OBJECTIVES	Code de couleurs des objectifs	Farbkode (für die Maßstabs-zahl) der Objektive
COHERENCE	Cohérence	Kohärenz
COHERENCE, DEGREE OF	Degré de cohérence	Kohärenzgrad
COHERENCE, PARTIAL	Cohérence partielle	partielle Kohärenz
COHERENCE CONDITION	Condition de cohérence	Kohärenzbedingung
COHERENT	Cohérent	kohärent
COLD CHAMBER	Chambre froid	Kühlkammer
COLLAR, CORRECTION	Bague de correction d'épaisseur de lamelle	Korrektionsfassung
COLLECTING POWER OF A LENS	Convergence d'une lentille	Sammelwirkung einer Linse
COLLECTOR	Lentille collectrice	Kollektor
COLLECTOR DIAPHRAGM	Diaphragme de lampe	Kollektorblende
COLLIMATE	Collimater	kollimieren
COLLIMATION	Collimation	Kollimation
COLLIMATOR	Collimateur	Kollimator
COLORIMETRY	Colorimétrie	Farbmessung
COLOUR	Couleur	Farbe
COLOUR, ACHROMATIC	Couleur achromatique perçue	unbunte Farbe
COLOUR, CHROMATIC	Couleur chromatique perçue	bunte Farbe
COLOUR, INTERFERENCE	Couleur d'interférence	Interferenzfarbe
COLOUR CODE OF OBJECTIVES	Code de couleurs des objectifs	Farbkode (für die Maßstabs-zahl) der Objektive
COLOUR-CONVERSION FILTER	Filtre de conversion de température de couleur	Farbkonversionsfilter
COLOUR MARKING OF OBJECTIVES	Marques de couleur des objectifs	Farbkennzeichnung (der speziellen Eigenschaften) der Objektive
COLOUR FILTER	Filtre coloré	Farbfilter
COLOUR SATURATION	Saturation d'une couleur	Farbsättigung
COLOUR TEMPERATURE	Température de couleur	Farbtemperatur
COLOUR VALUES, TRISTIMULUS	Composantes trichromatiques	Normfarbwerte; Farbmaßzahlen
COMA	Coma; Aigrette	Koma
COMPARISON MICROSCOPE	Microscope de comparaison	Vergleichsmikroskop
COMPENSATING EYEPIECE	Oculaire compensateur	Kompensokular
COMPENSATOR	Compensateur	Kompensator
COMPLEMENTARY DOMINANT WAVELENGTH	Longeur d'onde dominante complémentaire	kompensative Wellenlänge
COMPLEX RADIATION	Rayonnement polychromatique	zusammengesetzte Strahlung; Mischstrahlung
COMPOUND MICROSCOPE	Microscope (composé)	zusammengesetztes Mikroskop
COMPRESSARIUM	Chambre de compression	Kompressarium

APPENDIX II

CONDENSER	Condenseur	Kondensor
CONDENSER DIAPHRAGM	Diaphragme du condenseur	Kondensorblende
CONDENSER, ABBE	Condenseur d'Abbe	Abbescher Kondensor
CONDENSER, ACHROMATIC–APLANATIC	Condenseur aplanétique et achromatique	achromatisch-aplanatischer Kondensor
CONDENSER, CARDIOID	Condenseur cardioïde	Kardioidkondensor
CONDENSER, DARK GROUND	Condenseur pour fond noir	Dunkelfeldkondensor
CONDENSER, LAMP	Collecteur de lampe	Lampenkollektor
CONDENSER, PANCRATIC	Condenseur pancratique	pankratischer Kondensor
CONDENSER, PARABOLOID	Condenseur paraboloïde	Paraboloidkondensor
CONDENSER, PHASE-CONTRAST	Condenseur pour contraste de phase	Phasenkontrast-Kondensor
CONDENSER, SUBSTAGE	Condenseur sur sous-platine	Durchlichtkondensor
CONDENSER, SWING-OUT TOP LENS	Condenseur à lentille frontale escamotable	Kondensor mit ausklappbarer Frontlinse
CONDITION, COHERENCE	Condition de cohérence	Kohärenzbedingung
CONE, APLANATIC		aplanatischer Beleuchtungs-kegel
CONES (OF THE RETINA)	Cônes de la rétine	Zapfen (der Retina)
CONFOCAL IMAGING MODE	Imagerie confocale	konfokale Abbildung
CONFUSION, CIRCLE OF LEAST	Tache de moindre diffusion	Zerstreuungskreis
CONJUGATE	Conjugué	konjugiert
CONJUGATE PARTS OF A RAY		konjugierte Anteile eines Strahls
CONJUGATE PLANES	Plans conjugués	konjugierte Ebenen
CONJUGATE POINTS	Points conjugués	konjugierte Punkte
CONOSCOPIC (INTERFERENCE) FIGURE	Figure d'axes	konoskopische (Interferenz-) Figur
CONOSCOPIC OBSERVATION (OR CONOSCOPY)	Observation conoscopique (ou Conoscopie)	konoskopische Beobachtung (Konoskopie)
CONTINUOUS SPECTRUM	Spectre continu	koninuierliches Spektrum
CONTRAST	Contraste	Kontrast
CONTRAST, DIFFERENTIAL INTERFRENCE	Contraste interférentiel différentiel	differentieller Interferenz-kontrast
CONTRAST, INTERFERENCE	Contraste interférentiel	Interferenzkontrast
CONTRAST, MARGINAL	Contraste marginal	Randkontrast
CONTRAST, MODULATION	Contraste de modulation	Modulationskontrast
CONTRAST, PHOTOMETRIC	Contraste photométrique	photometrischer Kontrast
CONTRAST, PHYSIOLOGICAL	Contraste physiologique	physiologischer Kontrast
CONTRAST, RELIEF	Contraste de relief	Reliefkontrast
CONTRAST FILTER	Filtre de contraste	Kontrastfilter
CONVERGENCE	Convergence	Konvergenz
CONVERGENCE ANGLE	Angle de convergence	Konvergenzwinkel
CONVERGENCE POINT	Point de convergence	Konvergenzpunkt
CONVERGING LENS	Lentille convergente	Sammellinse
CONVERGING RAY BUNDLE	Faisceau de rayons convergents	konvergentes Strahlenbündel
COOLING STAGE	Platine réfrigérante	Kühltisch
CO-ORDINATES, CHROMATICITY	Coordonnées trichromatiques	Normfarbwertanteile; Normfarbwertkoordinaten
CORNEA	Cornée	Cornea; Hornhaut des Auges
CORRECTION	Correction	Korrektion
CORRECTION CLASS	Classe de correction	Korrektionsklasse
CORRECTION COLLAR	Bague de correction	Korrektionsfassung
CORRECTION FOR IMAGE DISTANCE	Distance image de correction	Korrektion auf die Bildweite
COUNTING CHAMBER	Cellule de comptage	Zählkammer
COUNTING EYEPIECE	Oculaire pour comptage	Zählokular
COVER GLASS (COVER SLIP)	Lamelle couvre-objet	Deckglas
CRITICAL ANGLE	Angle limite	kritischer Winkel
*CRITICAL ILLUMINATION	Eclairage critique	kritische Beleuchtung
CROSS, EXTINCTION	Croix noir	Auslöschungskreuz
CROSS HAIRS	Croisée de fils	Fadenkreuz
CROSS LINES	Réticule en croix	Strichkreuz
CROSS WIRES	Croisée de fils	Fadenkreuz; Strichkreuz
CROSSED POLARS	Polariseurs croisés	gekreuzte Polare (Polarisatoren)

CROWN GLASS	Crown	Kronglas
CRYSTAL OPTICAL AXIS	Axe optique d'un cristal	kristalloptische Achse
CRYSTALLINE LENS	Cristallin	Linse des Auges
CULTURE CHAMBER	Chambre de culture de tissus	(Zell-) Kulturkammer
CURVATURE OF IMAGE FIELD	Courbure de champ	Bildfeldkrümmung
CURVE, DISPERSION	Courbe de dispersion	Dispersionskurve
CURVE, EXTINCTION	Diagramme d'extinction	Extinktionskurve
CURVE, SPECTRAL LUMINOSITY	Courbe d'efficacité lumineuse relative spectrale de l'oeil, V (lamda)	spektrale Hellempfindlichkeitskurve; Vλ-Kurve
DARK ADAPTATION	Adaptation à l'obscurité	Dunkeladaptation
DARKFIELD	Fond noir	Dunkelfeld
DARKGROUND CONDENSER	Condenseur pour fond noir	Dunkelfeldkondensor
DARKGROUND ILLUMINATION	Éclairage en fond noir	Dunkelfeldbeleuchtung
DARKGROUND MICROSCOPY	Microscopie en fond noir	Dunkelfeldmikroskopie
DARKGROUND STOP	Diaphragme pour fond noir	Dunkelfeldblende
DAYLIGHT VISION	Vision photopique	Tagessehen
DEFINITION	Définition; Piqué (d'une image)	Bildschärfe; Auflösung
DEGREE OF COHERENCE	Degré de cohérence	Kohärenzgrad
DEGREE OF POLARIZATION	Taux de polarisation	Polarisationsgrad
DENSITY, OPTICAL	Densité optique	Extinktion
DEPTH OF FIELD (DEPTH OF SHARPNESS IN OBJECT SPACE)	Profondeur de champ	Schärfentiefe (im Objektraum)
DEPTH OF FOCUS (DEPTH OF SHARPNESS IN IMAGE SPACE)	Profondeur de foyer	Schärfentiefe (im Bildraum)
DIAGONAL POSITION	Orienté à 45 degrés; Position diagonale	Diagonalstellung
DIAGRAM, CHROMATICITY	Triangle des couleurs; Diagramme de chromaticité	Normfarbtafel; Farbdreieck
DIAPHRAGM	Diaphragme	Blende
DIAPHRAGM, APERTURE	Diaphragme d'ouverture	Aperturblende
DIAPHRAGM, BERTRAND	Diaphragme de Bertrand	Bertrandblende
DIAPHRAGM, COLLECTOR	Diaphragme de lampe	Kollektorblende
DIAPHRAGM, CONDENSER	Diaphragme du condenseur	Kondensorblende
DIAPHRAGM, FIELD	Diaphragme de champ	Feldblende
DIAPHRAGM, FIXED	Diaphragme fixe	feste Blende
DIAPHRAGM, ILLUMINATED FIELD	Diaphragme de champ	Leuchtfeldblende
DIAPHRAGM, ILLUMINATING APERTURE	Diaphragme d'ouverture	Beleuchtungsaperturblende
DIAPHRAGM, IRIS	Diaphragme iris	Irisblende
DIAPHRAGM, PHOTOMETRIC	Diaphragme photométrique	photometrische Blende; Photometerblende
DIAPHRAGM, PHOTOMICROGRAPHIC FIELD	Diaphragme du champ photographié	mikrophotographische Bildfeldblende
DIAPHRAGM, VISUAL FIELD	Diaphragme de champ de l'oculaire	Sehfeldblende
DICHROIC (DICHROMATIC) MIRROR	Miroir dichroïque	dichroitischer Spiegel; Farbteiler Spiegel
DICHROISM	Dichroïsme	Dichroismus
DIFFERENTIAL INTERFEROMETER	Interféromètre différentiel	differentielles Interferometer
DIFFERENTIAL INTERFERENCE CONTRAST	Contraste interférentiel différentiel	differentieller Interferenz-Kontrast
DIFFRACTED LIGHT	Lumière diffractée	gebeugtes Licht
DIFFRACTION	Diffraction	Beugung
DIFFRACTION, FRAUNHOFER	Diffraction de Fraunhofer	Fraunhofersche Beugung
DIFFRACTION, FRESNEL	Diffraction de Fresnel	Fresnelsche Beugung
DIFFRACTION APPARATUS, ABBE	Appareil de diffraction d'Abbe	Abbescher Diffraktionsapparat
DIFFRACTION DISC	Tache de diffraction	Beugungsscheibchen
DIFFRACTION GRATING	Réseau de diffraction	Beugungsgitter
DIFFRACTION LIMIT OF RESOLVING POWER	Résolution limitée par la diffraction	Begrenzung des Auflösungsvermögens durch Beugung
DIFFRACTION PATTERN	Figure de diffraction	Beugungsfigur

DIFFRACTION PATTERN (OF IMAGE), PRIMARY	Spectre de diffraction de l'objet; Spectre des fréquences spatiales	primäre Beugungsfigur; primäres Beugungsbild
DIMENSIONS, OPTICAL FITTING (OF THE MICROSCOPE)	Cotes optiques et mécaniques de référence du microscope	optische Anschlußmaße (des Mikroskops)
DIMENSIONS, OPTICAL FITTING, REFERENCE PLANE FOR	Plan de référence pour définir les cotes optiques et mécaniques du microscope	Bezugsebene für die optischen Anschlußmaße (des Mikroskops)
DIN	DIN	DIN
DIOPTRE	Dioptrie	Dioptrie
DIOPTRIC	Dioptrique	dioptrisch
DIPPING CONE	Cône d'immersion	Immersionsansatz-Kappe
DIRECT LIGHT	Lumière directe	direktes Licht
DIRECTION, EXTINCTION	Direction d'extinction	Auslöschungsrichtung
DIRECTION, VIBRATION	Direction de vibration	Schwingungsrichtung
DIRECTIONS, REFERENCE	Directions de référence	Bezugsrichtungen
DIRECTIONS, VIBRATION, PRINCIPAL	Directions principales de vibration	Hauptschwingungsrichtungen
DISC, DIFFRACTION	Tache d'Airy; tache de diffraction	Beugungsscheibchen
DISCHARGE LAMP	Lampe à décharge	(Gas-) Entladungslampe
DISPARITY	Disparité (de fixation)	Disparation
DISPERSION	Dispersion	Dispersion
DISPERSION CURVE	Courbe de dispersion	Dispersionskurve
DISPERSION STAINING MICROSCOPY	Micro-analyse par dispersion chromatique	Dispersionsfärbungs-Mikroskopie
DISPERSIVE POWER	Pouvoir dispersif	Zerstreuungsvermögen
DISSECTING MICROSCOPE	Microscope de dissection	Präpariermikroskop
DISTANCE, ACCOMMODATION	Distance d'accommodation	Akkommodationsentfernung
DISTANCE, FAR POINT	Distance du punctum remotum	Fernpunktsabstand
DISTANCE, FREE WORKING	Distance frontale	freier Arbeitsabstand
DISTANCE, IMAGE	Distance image	Bildweite
DISTANCE, INTERPUPILLARY	Écart d'yeux; écart inter-pupillaire	Pupillenabstand
DISTANCE, INTERSECTION	Distance frontale	Schnittweite
DISTANCE, MINIMUM RESOLVABLE	Limite de résolution	kleinster auflösbarer Abstand
DISTANCE, NEAR POINT	Distance du punctum proximum	Nahpunktabstand
DISTANCE, OBJECT	Distance objet	Objektweite
DISTANCE, OBJECT TO PRIMARY IMAGE	Distance entre l'objet et l'image intermédiaire	Objekt-Zwischenbild-Abstand
DISTANCE, OBJECTIVE TO PRIMARY IMAGE	Distance entre la face d'appui de l'objectif et l'image intermédiaire	Objektiv-Zwischenbild-Abstand
DISTANCE, OPTICAL	Chemin optique	optischer Abstand
DISTANCE, REFERENCE VIEWING	Distance conventionelle d'observation	Bezugssehweite
DISTANCE, RESOLVED	Distance résolue	aufgelöster Abstand
DISTANCES, CARDINAL		Kardinalstrecken
DISTORTION	Distorsion	Verzeichnung
DISTORTION, BARREL	Distorsion en barillet	tonnenförmige Verzeichnung
DISTORTION, PINCUSHION	Distorsion en coussinet	kissenförmige Verzeichnung
DISTRIBUTION TEMPERATURE	Température de répartition	Verteilungstemperatur
DIVERGENCE	Divergence	Divergenz
DIVERGENCE ANGLE	Angle de divergence	Divergenzwinkel
DIVERGING LENS	Lentille divergente	Zerstreuungslinse
DIVERGING RAY BUNDLE	Faisceau de rayons divergents	divergentes Strahlenbündel
DOMINANCE, OCULAR	Dominance oculaire	Äugigkeit; Führung eines Auges
DOMINANT WAVELENGTH	Longueur d'onde dominante	farbtongleiche Wellenlänge
DOUBLE-BEAM INTERFERENCE	Interférences à deux ondes	Zweistrahlinterferenz
DOUBLE-FOCUS INTERFEROMETER	Interéromètre a deux foyers	Doppelfokus-Interferometer
DOUBLE REFRACTION	Double réfraction	Doppelbrechung
DOUBLET	Doublet	Duplet
DRAWING APPARATUS	Appareil à dessiner	Zeichenapparat
DRAWING PRISM	Chambre claire	Zeichenprisma

DRAW TUBE	Tube coulissant	Ausziehtubus
DRY OBJECTIVE	Objectif à sec	Trockenobjektiv
EFFICACY (OF THE EYE), MAXIMUM SPECTRAL LUMINOUS	Efficacité lumineuse spectrale maximale	Maximalwert des spektralen photometrischen Strahlungsäquivalents
EFFICIENCY, LUMINOUS	Efficacité lumineuse relative	visueller Nutzeffekt
EFFICIENCY (OF THE EYE), SPECTRAL LUMINOUS	Efficacité lumineuse relative spectrale	spektraler Hellempfindlichkeitsgrad des Auges
ELECTRONIC FLASH	Flash électronique	Elektronenblitz
ELEMENTS, CARDINAL	Eléments cardinaux	Kardinalelemente
ELLIPTICALLY POLARIZED LIGHT	Lumière polarisée elliptique	elliptisch polarisiertes Licht
EMISSION	Emission	Emission; Abstrahlung
EMMETROPIA	Vision normale; Vision emmétrope	Emmetropie; Normalsichtigkeit
EMPTY MAGNIFICATION	Grossissement vide	Leervergrößerung
ENERGY, RADIANT	Énergie rayonnante	Strahlungsenergie
ENGLAND FINDER	Lame de England	
ENTOPTIC PHENOMENA	Phénomènes entoptiques	entoptische Erscheinungen
ENTRANCE PUPIL	Pupille d'entrée	Eintrittspupille
ENTRANCE PUPIL OF THE EYE	Pupille d'entrée de l'oeil	Eintrittspupille des Auges
ENVELOPE	Ballon; Ampoule	Lampenkolben
EPI-ILLUMINATION	Éclairage en lumière incidente	Auflichtbeleuchtung
EXCITATION	Excitation	Anregung
EXCITER FILTER	Filtre d'excitation	Erregerfilter
EXITANCE, LUMINOUS (AT A POINT ON A SURFACE)	Exitance lumineuse (en un point d'une surface)	spezifische Lichtausstrahlung (an einem Punkt einer Fläche)
EXITANCE, RADIANT (AT A POINT ON A SURFACE)	Exitance énergétique (en un point d'une surface)	spezifische Ausstrahlung (an einem Punkt einer Fläche)
EXIT PUPIL	Pupille de sortie	Austrittspupille
EXIT PUPIL OF THE MICROSCOPE	Pupile de sortie du microscope	Austrittspupille des Mikroskops
EXPOSURE	Lumination	Belichtungsvorgang
EXPOSURE, LIGHT	Exposition lumineuse; Lumination	Belichtung
EXPOSURE METER	Posemètre	Belichtungsmesser
EXPOSURE, RADIANT (AT A POINT OF A SURFACE)	Exposition énergétique (en un point d'une surface)	Bestrahlung (an einem Punkt einer Fläche)
EXTENDED SOURCE	Source étendue	ausgedehnte Lichtquelle
EXTERNAL-DIAPHRAGM EYEPIECE	Oculaire positif	Okular mit Vorderblende
EXTINCTION	Extinction	Auslöschung
EXTINCTION, OBLIQUE	Extinction oblique	schiefe Auslöschung
EXTINCTION, STRAIGHT	Extinction droite	gerade Auslöschung
EXTINCTION, SYMMETRICAL	Extinction symétrique	symmetrische Auslöschung
EXTINCTION ANGLE	Angle d'extinction	Auslöschungswinkel
EXTINCTION CROSS	Croix noire	Auslöschungskreuz
EXTINCTION CURVE	Diagramme d'extinction	Auslöschungskurve
EXTINCTION DIRECTION	Direction d'extinction	Auslöschungsrichtung
EXTINCTION FACTOR	Facteur d'extinction	Extinktionsfaktor
EXTINCTION POSITION	Position d'extinction	Auslöschungsstellung
EXTRAORDINARY RAY	Rayon extrordinaire	außerordentlicher Strahl
EYE	Oeil	Auge
EYE, LENS OF	Cristallin	Linse des Auges
EYE LENS	Verre d'oeil	Augenlinse
EYEPIECE	Oculaire	Okular
EYEPIECE, COMPARISON	Tube de comparaison	Vergleichsokular
EYEPIECE, COMPENSATING	Oculaire compensateur	Kompensokular; Kompensationsokular
EYEPIECE, COUNTING	Oculaire de comptage	Zählokular
EYEPIECE, DEMONSTRATION	Tube de démonstration	Doppelokular
EYEPIECE, EXTERNAL-DIAPHRAGM	Oculaire positif	Okular mit Vorderblende
EYEPIECE, FLAT FIELD	Oculaire à champ plan	Planokular

EYEPIECE, FOCUSABLE	Oculaire à mise au point réglable	einstellbares Okular
EYEPIECE, GONIOMETER	Oculaire goniométrique	Goniometerokular
EYEPIECE, GRATICULE	Oculaire à graticule	Strichplattenokular
EYEPIECE, HIGH-EYEPOINT	Oculaire pour porteurs de lunettes	Brillenträgerokular
EYEPIECE, HOMAL®	Homal®	Homal®
EYEPIECE, HUYGENIAN	Oculaire de Huygens	Huygensokular
EYEPIECE, IMAGE SHEARING	Oculaire à dédoublement d'images	Doppelbild-Okular
EYEPIECE, INTEGRATING	Oculaire de comptage	Integrationsokular
EYEPIECE, INTERNAL-DIAPHRAGM	Oculaire négatif	Okular mit Zwischenblende
EYEPIECE, KELLNER	Oculaire de Kellner	Kellnersches Okular
EYEPIECE, LOCATING FLANGE OF	épaulement de l'oculaire	Anlagefläche des Okulars
EYEPIECE, MICROMETER	Oculaire micrométrique	Mikrometerokular
EYEPIECE, MICROMETER-SCREW	Oculaire à vis micrométrique	Okularschraubenmikrometer
EYEPIECE, NEGATIVE	Oculaire négatif	
EYEPIECE, ORTHOSCOPIC	Oculaire orthoscopique; (de Kellner)	orthoskopisches Okular
EYEPIECE, PARFOCALIZING DISTANCE OF	Distance d'équilibrage de l'oculaire	Abgleichlänge des Okulars
EYEPIECE, POINTER	Oculaire à aiguille indicatrice	Zeigerokular
EYEPIECE, POSITIVE	Oculaire positif	
EYEPIECE, PROJECTION	Oculaire de projection	Projektionsokular
EYEPIECE, RAMSDEN	Oculaire de Ramsden	Ramsden-Okular
EYEPIECE, SLOTTED	Oculaire à tirette de Wright	Wrightsches Okular
EYEPIECE, WIDE-FIELD	Oculaire à grand champ	Großfeldokular
EYEPIECE-GRATICULE	Graticule oculaire	Okular-Strichplatte
EYEPIECE-LOCATING SURFACE (OF VIEWING TUBE)	Face d'appui de l'oculaire (sur le tube)	Okular-Anlagefläche des Beobachtungstubus
EYEPOINT	Anneau oculaire; Anneau de Ramsden	Augenkreis; Ramsdenscher Kreis
EYEPOINT HEIGHT	Distance d'anneau oculaire	Höhe der Austrittspupille
FACTOR, EXTINCTION	Facteur d'éxtinction	Extinktionsfaktor
FACTOR, LUMINANCE	Facteur de lumninance	Hellbezugswert
FAR-POINT DISTANCE	Distance du punctum remotum	Fernpunktsabstand
FAR-POINT OF THE EYE	Punctum remotum	Fernpunkt des Auges
FARSIGHTEDNESS	Hypermétropie	Weitsichtigkeit; Hypermetropie
FIBRE OPTIC	Faisceau de fibres optiques	Faseroptik
FIBRE OPTIC ILLUMINATOR	Illuminateur à fibres optiques	Faseroptik-Leuchte
FIELD	Champ	Feld
FIELD, DEPTH OF	Profondeur de champ	Schärfentiefe
FIELD, ILLUMINATED	Champ éclairé	Leuchtfeld
FIELD, IMAGE	Champ image	Bildfeld
FIELD, OBJECT	Champ objet	Objektfeld
FIELD, PHOTOMETRIC	Champ photométré	Photometerfeld
FIELD, SURROUNDING	Champ périphérique	Umfeld
FIELD, VISUAL	Champ visuel	Gesichtsfeld
FIELD DIAPHRAGM	Diaphragme de champ	Feldblende
FIELD LENS	Lentille de champ	Feldlinse
FIELD MICROSCOPE	Microscope de voyage	Reisemikroskop
FIELD OF VIEW	Champ d'observation	Sehfeld
FIELD-OF-VIEW NUMBER	Indice de champ	Sehfeldzahl
FIELD PLANE	Champ (objet ou image)	Feldebene
FIGURE, CONOSCOPIC	Figure conoscopique	konoskopisches Bild
FIGURE, POLARIZATION	Figure de convergence	Achsenbild
FILAMENT	Filament	Glühwendel
FILAMENT-LAMP	Lampe à incandescence	Glühfadenlampe; (Glühbirne)
FILAR MICROMETER	Oculaire à vis micrométrique	Okularschraubenmikrometer
FILTER	Filtre	Filter

FILTER, BARRIER	Filtre de blockage; Filtre d'arrêt	Sperrfilter
FILTER, BROAD-BAND-PASS (OR BROAD-BAND)	Filtre à large bande passante	Breitbandfilter
FILTER, COLOR	Filtre coloré	Farbfilter
FILTER, COLOR-CONVERSION (OR CONVERSION)	Filtre de conversion de température de couleur	Farbkonversionsfilter; Konversionsfilter
FILTER, CONTRAST	Filtre coloré pour modifier le contraste	Kontrastfilter
FILTER, EXCITER	Filtre d'excitation	Erregerfilter
FILTER, HEAT (OR HEAT-PROTECTION)	Filtre anti-calorique	Wärmeschutzfilter
FILTER, INTERFERENCE	Filtre interférentiel	Interferenzfilter
FILTER, LONG-WAVE-PASS (OR LONG-PASS)	Filtre passe haut	Langpaßfilter
FILTER, NARROW-BAND-PASS (OR NARROW-BAND)	Filtre à bande étroite	Schmalbandfilter
FILTER, NEUTRAL DENSITY	Filtre gris; Densité neutre	Graufilter; Neutralfilter
FILTER, POLARIZING	Filtre polarisant	Polarisationsfilter
FILTER, SHORT-WAVE-PASS	Filtre pass-bas	Kurzpaßfilter
FILTER TRAY	Support de filtre	Filterhalter
FINDER	Lame de repérage	Objektfinder
FINE ADJUSTMENT	Mouvement lent	Feintrieb
FIRST-ORDER RED	Pourpre teinte sensible du 1° ordre	Rot 1. Ordnung
FIRST-ORDER RED PLATE	Lame à teinte sensible du 1° ordre	Rot 1. Ordnung-Plättchen
FIXATION	Fixation	Fixieren
FIXATION LINE	Ligne principale de visée	Fixierlinie
FIXATION POINT	Point de fixation	Fixierpunkt
FIXED DIAPHRAGM	Diaphragme fixe	feste Blende
FLASH, ELECTRONIC	Flash électronique	Elektronenblitz
FLAT-FIELD EYEPIECE	Oculaire à champ plan	Planokular
FLAT-FIELD OBJECTIVE	Objectif à champ plan	Planobjektiv
FLINT GLASS	Flint	Flintglas
FLUORESCENCE	Fluorescence	Fluoreszenz
FLUORESCENCE, AUTO	Fluorescence propre (primaire)	Eigenfluoreszenz
FLUORESCENCE MICROSCOPE	Microscope pour fluorescence	Fluoreszenzmikroskop
FLUORESCENCE MICROSCOPY	Microscopie par fluorescence	Fluoreszenzmikroskopie
FLUORESCENCE, PRIMARY	Fluorescence propre (primaire)	Eigenfluoreszenz
FLUORESCENCE, SECONDARY	Fluorescence induite (secondaire)	Sekundärfluoreszenz
FLUORIMETRY	Fluorimétrie	Fluorometrie
FLUORIMETRY, MICROSCOPE	Microfluorimétrie	Mikroskop-Fluorometrie
FLUORITE	Fluorine	Fluorit; Flußspat
*FLUORITE OBJECTIVE	*Objectif à la fluorine	*Fluoritobjektiv
FLUOROCHROME	Fluorochrome	Fluorochrom; Fluoreszenz-Farbstoff
FLUOROPHORE	Fluorophore	Fluorophor
FLUX, LUMINOUS	Flux lumineux	Lichtstrom
FLUX, RADIANT	Flux énergétique	Strahlungsfluß
FOCAL DEPTH	Profondeur de foyer	Schärfentiefe
FOCAL LENGTH	Distance focale	Brennweite
FOCAL POINT	Foyer	Brennpunkt
FOCAL PLANE	Plan focal	Brennebene
FOCAL PLANE, BACK	Plan focal image	hintere Brennebene
FOCAL PLANE, FRONT	Plan focal objet	vordere Brennebene
FOCUS	Foyer	Brennpunkt
FOCUSABLE EYEPIECE	Oculaire à mise au point réglable	einstellbares Okular
FOCUSING MAGNIFIER	Loupe de mise au point	Einstell-Lupe
FOCUSING MECHANISM (OF THE MICROSCOPE)	Mouvement lent (du microscope)	Einstellmechanismus (des Mikroskops)
FOCUSING SCREEN, CLEAR	Verre clair de mise au point	Einstell-Klarscheibe

FOVEA	Fovea	Fovea (centralis); Netzhautgrube
FRAUNHOFER DIFFRACTION	Diffraction de Fraunhofer	Fraunhofersche Beugung
FREE WORKING DISTANCE (OF THE OBJECTIVE)	Distance frontale (de l'objectif	freier Arbeitsabstand (des Objektivs)
FRESNEL DIFFRACTION	Diffraction de Fresnel	Fresnelsche Beugung
FRESNEL LENS	Lentille de Fresnel	Fresnel-Linse
FRONT FOCAL PLANE	Plan focal objet	vordere Brennebene
FRONT LENS	Lentille frontale	Frontlinse
GAUSSIAN SPACE	Espace de Gauss	Gaußscher Raum
GEOMETRICAL OPTICS	Optique géométrique	geometrische Optik
GLARE	Eblouissement; Lumière parasite	Blendung; Überstrahlung; Falschlicht
GLASS, CROWN	Crown	Kronglas
GLASS, FLINT	Flint	Flintglas
GLASS, GROUND	Verre dépoli	Mattscheibe
GLASS, HEAT ABSORBING	Verre anti-calorique	Wärmeschutzglas
GLASS, OPAL	Verre opale	Opalglas
GONIOMETER	Goniomètre	Goniometer; Winkelmesser
GONIOMETER EYEPIECE	Oculaire goniométrique	Goniometer-Okular
GRATICULE	Graticule	Strichplatte
GRATICULE, EYEPIECE	Graticule oculaire	Okularstrichplatte
GRATICULE EYEPIECE	Oculaire à graticule	Strichplattenokular; Meßokular
GRATING, DIFFRACTION	Réseau de diffraction	Beugungsgitter
GREENOUGH MICROSCOPE	Microscope de Greenough	Greenough-Mikroskop
GROUND GLASS	Verre dépoli	Mattscheibe; Mattglas
HAEMOCYTOMETER	Hématimètre; Cellule pour numération sanguine	Blutkörperchen-Zählkammer
HALF-PEAK-HEIGHT BANDWIDTH	Bande passante à mi-hauteur	Halbwertsbreite
HALO	Halo	Halo
HALOGEN LAMP	Lampe à halogène	Halogenlampe
HANGING-DROP	Goutte pendante	hängender Tropfen
HANGING DROP SLIDE	Lame creuse	Hohlschliff-Objektträger
HEAD, CAMERA	Tête photographique	Photokopf
HEAD, PHOTOMETER	Tête photométrique	Photometerkopf
HEAD, PROJECTION	Tête de projection	Projektionskopf
HEAD, TELEVISION	Tête pour caméra de télévision	Fernsehkopf
HEAD, TUBE	Raccord photographique	Tubuskopf
HEAT-ABSORBING GLASS	Verre anti-calorique	Wärmeschutzglas
HEAT FILTER	Filtre anti-calorique	Wärmeschutzfilter
HEATING STAGE	Platine chauffante	Heiztisch
HIATUS	Interstice	Interstitium
HIGH-EYEPOINT EYEPIECE	Oculaire pour porteurs de lunettes	Brillenträgerokular
HOMAL®	Homal®	Homal®
HOMOGENEOUS IMMERSION	Immersion homogène	homogene Immersion
HUE	Teinte	Farbton
HUYGENIAN EYEPIECE	Oculaire de Huygens	Huygens-Okular
HYPERMETROPIA	Hypermétropie	Hypermetropie; Weitsichtigkeit
ILLUMINANCE (AT A POINT ON A SURFACE)	Éclairement lumineux (en un point d'une surface)	Beleuchtungsstärke (an einem Punkt einer Fläche)
ILLUMINATED FIELD	Champ éclairé	Leuchtfeld
ILLUMINATED FIELD DIAPHRAGM	Diaphragme de champ	Leuchtfeldblende
ILLUMINATING APERTURE DIAPHRAGM	Diaphragme d'ouverture du condenseur	Beleuchtungsaperturblende
ILLUMINATING SYSTEM	Système d'éclairage	Beleuchtungseinrichtung
ILLUMINATION	Éclairage	Beleuchtung
ILLUMINATION, ANNULAR	Éclairage annulaire	ringförmige Beleuchtung
ILLUMINATION, AXIAL	Éclairage axial	axiale Beleuchtung
ILLUMINATION, BRIGHT FIELD	Éclairage en fond clair	Hellfeldbeleuchtung

*ILLUMINATION, CRITICAL	Éclairage critique	kritische Beleuchtung
ILLUMINATION, DARK-GROUND	Éclairage en fond noir	Dunkelfeldbeleuchtung
ILLUMINATION, EPI-	Éclairage épiscopique	Auflichtbeleuchtung
ILLUMINATION, INCIDENT	Éclairage en lumière incidente	auffallende (einfallende) Beleuchtung
ILLUMINATION, KÖHLER	Éclairage de Köhler	Köhlersche Beleuchtung
ILLUMINATION, NELSONIAN	Éclairage de Nelson	
ILLUMINATION, RHEINBERG	Éclairage de Rheinberg	
ILLUMINATION, SOURCE-FOCUSED	Éclairage critique	kritische Beleuchtung
ILLUMINATION, TRANSMITTED-LIGHT (OR TRANS-ILLUMINATION)	Éclairage en lumière transmise	Durchlichtbeleuchtung
ILLUMINATION, UNILATERAL OBLIQUE	Éclairage oblique	einseitig schiefe Beleuchtung
ILLUMINATION, VERTICAL	Éclairage en lumière incidente	(axiale) Auflichtbeleuchtung
ILLUMINATOR, EPI-	Illuminateur vertical	Auflichtilluminator
ILLUMINATOR, FIBRE OPTIC	Illuminateur à fibre optiques	Faseroptik-Leuchte
IMAGE	Image	Bild
IMAGE, AERIAL	Image aérienne	Luftbild
IMAGE, BRIGHT FIELD	Image en fond clair	Hellfeldbild
IMAGE, DARK GROUND	Image en fond noir	Dunkelfeldbild
IMAGE, PRIMARY	Image intermédiaire	reelles Zwischenbild
IMAGE, REAL	Image réelle	reelles Bild
IMAGE, RETINAL	Image rétienne	Netzhautbild
IMAGE, VIRTUAL	Image virtuelle	virtuelles Bild
IMAGE ANALYSIS	Analyse d'image	Bildanalyse
IMAGE CONVERTER	Convertisseur d'image	Bildwandler
IMAGE DISTANCE	Distance d'image	Bildweite
IMAGE DISTANCE CORRECTION OF OBJECTIVE	Distance de correction d'un objectif	Bildweitenkorrektion des Objektivs
IMAGE FIELD	Champ image	Bildfeld
IMAGE FIELD, CURVATURE OF	Courbure de champ image	Bildfeldkrümmung
IMAGE INTENSIFIER	Amplificateur de luminances	Bildverstärker
IMAGE PLANE	Plan image	Bildebene
IMAGE-SIDE APERTURE	Ouverture numérique image	bildseitige numerische Apertur
IMAGE SPACE	Espace image	Bildraum
IMAGING DEPTH	Profondeur de champ image	Abbildungstiefe
IMAGING MODE, CONFOCAL	Imagerie confocale	konfokale Abbildung
IMAGING SCALE	Grandissement	Abbildungsmaßstab
IMMERSION	Immersion	Immersion
IMMERSION, HOMOGENEOUS	Immersion homogène	homogene Immersion
IMMERSION, HOMOGENEOUS OF OBJECTIVE (OR CONDENSER)	Immersion homogène de l'objectif (ou du condenseur)	homogene Immersion des Objektivs (oder des Kondensors)
IMMERSION LENS	Optique à immersion	Immersionssystem
IMMERSION LIQUID	Liquide d'immersion	Immersionsflüssigkeit
IMMERSION OIL	Huile d'immersion	Immersionsöl
INCANDESCENCE	Incandescence	Glühen
INCIDENCE	Incidence	Einfall; Auffall
INCIDENCE, ANGLE OF	Angle d'incidence	Einfallswinkel; Auffallswinkel
INCIDENT BEAM	Faisceau incident	einfallendes (oder auf-fallendes) Licht- (oder Strahlen-) Bündel
INCIDENT ILLUMINATION	Éclairage épiscopique	Auflichtbeleuchtung
INCIDENT RAY	Rayon incident	einfallender Strahl
INCOHERENT	Incohérent	inkohärent
INDEX OF REFRACTION	Indice de réfraction	Brechungszahl
INDICATRIX	Ellipsoïde des indices	Indikatrix
INFINITY-CORRECTED OBJECTIVE	Objectif corrigé à l'infini	auf unendlich korrigiertes Objektiv
INFRA-RED MICROSCOPY	Microscopie dans l'infra-rouge	Infrarotmikroskopie
INFRA-RED RADIATION	Rayonnement infra-rouge	Infrarotstrahlung

APPENDIX II

INTEGRATING EYEPIECE	Oculaire de comptage	Integrationsokular
INTEGRATING STAGE	Platine de comptage	Integrationstisch
INTENSITY	Intensité	Intensität; Stärke
INTENSITY, LUMINOUS	Intensité lumineuse	Lichtstärke
INTENSITY, RADIANT	Intensité énergétique	Strahlstärke
INTENSITY CONTRAST	Contraste	Intensitätskontrast; photometrischer Kontrast
INTERFERENCE	Interférences	Interferenz
INTERFERENCE, DOUBLE-BEAM	Interférences à deux ondes	Zweistrahlinterferenz
INTERFERENCE COLOUR	Couleurs interférentielles	Interferenzfarbe
INTERFERENCE CONTRAST	Contraste interférentiel	Interferenzkontrast
INTERFERENCE FILTER	Filtre interférentiel	Interferenzfilter
INTERFERENCE IMAGE, PRIMARY	Spectre de diffraction de l'objet; spectre des fréquences spatiales	primäres Interferenzbild
INTERFERENCE MICROSCOPE	Micro-interféromètre	Interferenzmikroskop
INTERFERENCE MICROSCOPY	Microscopie interférentielle	Interferenzmikroskopie
INTERFERENCE, MULTIPLE BEAM	Interférences à ondes multiples	Vielstrahlinterferenz
INTERFERENCE, POLARIZING	Interférences, par double refraction, en lumière polarisée	Polarisationsinterferenz
INTERFEROMETER, DIFFERENTIAL	Interféromètre différentiel	differentielles Interferometer
INTERFEROMETER,DOUBLE-FOCUS	Interféromètre à deux foyers	Doppelfokus-Interferometer
INTERFEROMETER, SHEARING	Interféromètre à dédoublement lateral	Shearing-Interferometer
INTERFEROMETRY	Interférométrie	Interferometrie
INTERFEROMETRY, MICROSCOPE	Micro-interférométrie	Mikrointerferometrie
INTERMEDIATE LENS	Optique de tube	Tubuslinse; Zwischenlinse
INTERMEDIATE TUBE	Tube intermédiaire	Zwischentubus
INTERNAL-DIAPHRAGM EYEPIECE	Oculaire négatif	Okular mit Zwischenblende
INTERNAL TRANSMISSION DENSITY	Densité optique interne	Extinktion
INTERPUPILLARY DISTANCE	Écart d'yeux; écart interpupillaire	Pupillenabstand
INTERSECTION DISTANCE	Distance frontale	Schnittweite
INVERSION	Inversion de pseudo-relief	Inversion
INVERTED MICROSCOPE	Microscope inversé; Microscope de le Chatelier	umgekehrtes Mikroskop
IRIS	Iris	Iris (des Auges); Regenbogenhaut
IRIS DIAPHRAGM	Diaphragme iris	Irisblende
IRRADIANCE	Eclairement énergétique	Bestrahlungsstärke
IRRADATION	Irradation	Bestrahlung
ISO	ISO	ISO
ISOCHROMATIC CURVES	Lignes isochromatiques	Isochromaten
ISOGYRES	Isogyres	Isogyren
ISOTROPY	Isotropie	Isotropie
KELLNER EYEPIECE	Oculaire de Kellner	Kellnersches Okular
KÖHLER ILLUMINATION	Éclairage de Köhler	Köhlersche Beleuchtung
LAMP	Lampe	Lampe
LAMP, DISCHARGE	Lampe à décharge	(Gas-) Entladungslampe
LAMP, FILAMENT	Lampe à incandescence	Glühlampe
LAMP, HALOGEN	Lampe à halogène	Halogenlampe
LAMP, MERCURY ARC	Lampe à vapeur de mercure	Quecksilberdampflampe
LAMP, MICROSCOPE	Lampe de microscope	Mikroskopierleuchte
LAMP, SOLID-SOURCE		
LAMP, XENON ARC	Lampe à xénon	Xenonlampe
LASER	Laser	Laser
LATERAL CHROMATIC ABERRATION	Chromatisme latéral	chromatische Queraberration; chromatische Vergrößerungsdifferenz
LATERAL MAGNIFICATION	Grandissment transversal	Lateralvergrößerung; Abbildungsmaßstab

LAW, BEER–LAMBERT	Loi de Beer-Lambert	Lambert–Beersches Gesetz
LAW, SNELL'S	Loi de Descartes	Snelliussches Brechungsgesetz
LEMNISCATE	Lemniscate	Lemniskate
LENS	Lentille; Objectif	Linse
LENS, ASPHERICAL	Lentille asphérique	asphärische Linse
LENS, BERTAND (AMICI–BERTRAND)	Lentille de Bertrand (Amici–Bertrand)	Amici–Bertrand-Linse; Bertrand-Linse
LENS, COLLECTOR	Collecteur de lampe	Kollektor
LENS, CONVERGING	Lentille convergente	Sammellinse
LENS, DIVERGING	Lentille divergente	Zerstreuungslinse
LENS, EYE	Lentille d'oeil	Augenlinse
LENS, FIELD	Lentille de champ	Feldlinse
LENS, FRONT	Lentille frontale	Frontlinse
LENS, IMMERSION	Objectif à immersion	Immersionssystem
LENS, INTERMEDIATE	Système optique intermédiaire	Zwischenlinse
LENS, MENISCUS	Lentille ménisque	Meniskus; Meniskuslinse
LENS, NEGATIVE	Système optique divergent	Negativlinse
LENS, POSITIVE	Système optique convergent	Positivlinse
LENS, RELAY	Véhicule	Übertragungslinse, -system
LENS, THICK	Lentille épaisse	dicke Linse
LENS, THIN	Lentille mince	dünne Linse
LENS, TOP	Lentille frontale du condenseur	Frontlinse des Kondensors
LENS, TUBE-LENGTH CORRECTION	Lentille de modification de la longueur de tube	Tubuslängen-Korrekturlinse
LENS, OF THE EYE	Cristallin	Linse des Auges
LENS MOUNT	Monture de lentille	Linsenfassung
LEVELLING (OF A POLISHED SECTION)	Montage d'un échantillon poli &	Ausrichten (eines polierten Anschliffs)
LEVELLING STAGE	Platine orientable	Anschlifftisch
LIEBERKÜHN (LIEBERKÜHN REFLECTOR)	Miroir de Lieberkühn	Lieberkühnspiegel
LIGHT (VISIBLE RADIATION)	Lumière (Rayonnement visible)	Licht (sichtbare Strahlung)
LIGHT, CIRCULARY-POLARIZED	Lumière polarisée circulaire	zirkular polarisiertes Licht
LIGHT, DIFFRACTED	Lumière diffractée	abgebeugtes Licht
LIGHT, DIRECT	Lumière directe	direktes Licht
LIGHT, ELLIPTICALLY-POLARIZED	Lumière polarisée elliptique	elliptisch polarisiertes Licht
LIGHT, NATURAL	Lumière naturelle	natürliches Licht
LIGHT, PLANE-POLARTIZED	Lumière polarisée rectiligne	linear polarisiertes Licht
LIGHT, POLARIZED	Lumière polarisée	polarisiertes Licht
LIGHT, RING	Source circulaire	Ringleuchte
LIGHT, STRAY	Lumière parasite	Streulicht
LIGHT, WHITE	Lumière blanche	weißes Licht
LIGHT ADAPTATION	Adaptation à la lumière	Helladaptation
LIGHT ADAPTATION, STATE OF	État d'adaptation à la lumière	Helladaptationszustand
LIGHT BEAM	Faisceaux lumineux	Lichtbündel
LIGHT EXPOSURE	Exposition lumineuse; Lumination	Belichtung
LIGHT FILTER	Filtre	Lichtfilter
LIGHT MICROSCOPY	Microscopie (optique)	Lichtmikroskopie
LIGHT PATH	Chemin optique	Strahlengang; Lichtweg
LIGHT SOURCE	Source lumineuse	Lichtquelle
LIGHT VECTOR	Vecteur électrique; Vecteur de Fresnel; Vibration lumineuse	Lichtvektor
LIMB	Potence	Tubusträger; Stativbügel
LINE, FIXATION	Ligne principale de visée	Fixierlinie
LINE SPECTRUM	Spectre de raies	Linienspektrum
LINEAR MAGNIFICATION	Grandissement latéral	Linearvergrößerung; Abbildungsmaßstab
LOCATING SURFACE (OR FLANGE)	Face d'appui	Anlagefläche
LOCATING FLANGE OF EYEPIECE	Épaulement de l'oculaire	Anlagefläche des Okulars

LOCATING FLANGE OF OBJECTIVE	Épaulement de l'objectif	Anlagefläche des Objektivs
LOCATING SURFACE FOR EYEPIECE	Face d'appui de l'oculaire	Anlagefläche für das Okular
LOCATING SURFACE FOR OBJECTIVE (OF THE NOSEPIECE)	Face d'appui de l'objectif (*sur le revolver*)	Anlagefläche für das Objektiv (am Objektiv-revolver)
LOCUS, SPECTRUM	Spectrum locus	Farbort
LONG-WAVE-PASS FILTER	Filtre passe-haut	Langpaßfilter
LONGITUDINAL MAGNIFICATION	Grandissement longitudinal; (axial)	Axialvergrößerung
LOUPE	Loupe	Lupe
LUMEN	Lumen	Lumen
LUMINANCE	Luminance	Leuchtdichte
LUMINANCE FACTOR	Facteur de luminance	Hellbezugswert
LUMINESCENCE	Luminescence	Lumineszenz
LUMINOSITY	Luminosité	Helligkeit
LUMINOSTIY CURVE	Courbe d'efficacité lumineuse relative spectrale de l'oeil, V(lamda)	Hellempfindlichkeitskurve des Auges; Vλ-Kurve
LUMINOUS EFFICIENCY	Efficacité lumineuse relative	visueller Nutzeffekt
LUMINOUS EXITANCE	Exitance lumineuse	spezifische Lichtaus-strahlung
LUMINOUS FLUX	Flux lumineux	Lichtstrom
LUMINOUS INTENSITY	Intensité lumineuse	Lichtstärke
LUMINOUS TRANSMITTANCE	Facteur de transmission lumineuse	Helltransmissionsgrad
LUX	Lux	Lux
MACROPHOTOGRAPHY	Macrophotographie	Makrophotographie
MAGNIFICATION	Grandissment	Vergrößerung
MAGNIFICATION, ANGULAR	Grossissement commercial	Angularvergrößerung
MAGNIFICATION, AREAL		Flächenvergrößerung
MAGNIFICATION, AXIAL	Grandissemant longitudinal; axial	Axialvergrößerung
MAGNIFICATION, EMPTY	Grossissement vide	leere Vergößerung
MAGNIFICATION, LATERAL	Grandissement transversal	Lateralvergrößerung
MAGNIFICATION, LINEAR	Grandissement transversal	Linearvergrößerung
MAGNIFICATION, PHOTOMICROGRAPHIC	Grandissement photographique; Échelle	mikrophotographische Vergrößerung
MAGNIFICATION, PRIMARY	Grandissement de l'objectif	Maßstabszahl des Objektivs
MAGNIFICATION, TOTAL	Grossissement total du microscope	Gesamtvergrößerung
MAGNIFICATION, USEFUL RANGE OF	Domaine des grossissements utiles	Bereich der förderlichen Vergrößerung
MAGNIFICATION CHANGER	Changeur de grossissement	Vergrößerungswechsler
MAGNIFICATION NUMBER OF OBJECTIVE	Grandissement de l'objectif	Maßstabszahl des Objektivs
MAGNIFIER	Loupe	Lupe
MAGNIFIER, FOCUSING	Loupe de mise au point	Einstell-Lupe
MAGNIFYING POWER	1)Grandissement (if the lens gives a real image) 2)Grossissement commercial (if the lens gives a virtual image for visual observation at infinity)	Vergrößerungsvermögen; Vergrößerung
MAGNIFYING POWER OF EYEPIECE	Grossissement commercial (ou conventionnel) de l'oculaire	Okularvergrößerung
MAGNIFYING POWER OF OBJECTIVE	Grandissement de l'objectif	Maßstabszahl des Objektivs
MARGINAL CONTRAST	Contraste marginal	Randkontrast
MARGINAL RAY	Rayon marginal	Randstrahl
MARKER, OBJECT	Marqueur d'objets	Objektmarkierer
MARKING (OF OPTICAL COMPONENTS)	Marquage (des pièces optiques)	Beschriftung optischer Teile

MARKING, COLOUR, OF OBJECTIVES	Identification des objectifs par marques de couleurs	Farbkennzeichnung (der speziellen Eigenschaften) von Objektiven
MASK	Cadre	Bildmaske
MAXIMUM SPECTRAL LUMINOUS EFFICACY (OF THE EYE)	Efficacité lumineuse spectrale maximale	Maximalwert des spektralen photometrischen Strahlungs-äquivalents
MEASURING MICROSCOPE	Microscope de mesures	Meßmikroskop
MECHANICAL STAGE	Platine à mouvements croisés; Surplatine	Kreuztisch; Objektführer
MECHANICAL TUBELENGTH	Longeur (mécanique) de tube	mechanische Tubuslänge
MECHANISM, FOCUSING	Mouvement de mise au point	Fokussierungsmechanismus
MECHANISM, FOCUSING (OF THE MICROSCOPE)	Mouvement de mise au point du microscope	Fokussierungsmechanismus des Mikroskops
MEDIUM	Milieu	Medium
MERCURY ARC LAMP	Lampe à vapeur de mercure	Quecksilberdampflampe
MESOPIC VISION	Vision mésopique	mesopisches Sehen; Dämmerungssehen
METALLOGRAPHY	Métallographie	Metallographie
METER, EXPOSURE	Posemètre	Belichtungsmesser
MICRODENSITOMETER	Microdensitomètre	Mikrodensitometer
MICROGRAPH	Photomicrographie	mikrophotographische Aufnahme
MICROHARDNESS TESTER	Microduromètre	Mikrohärteprüfer
MICROINTERFEROMETER	Microinterféromètre	Mikrointerferometer
MICROMANIPULATOR	Micromanipulateur	Mikromanipulator
MICROMETER	Micromètre	Mikrometer
MICROMETER, FILAR	Oculaire à vis micrométrique	Okularschraubenmikrometer
MICROMETER STAGE	Micromètre-objet	Objektmikrometer
MICROMETER EYEPIECE	Oculaire micrométrique	Mikrometerokular; Meßokular
MICROMETER-SCREW EYEPIECE	Oculaire à vis micrométrique	Okularschraubenmikrometer
MICROMETRE	Micromètre, μm	Mikrometer, μm
MICRON	Micron, μm	Mikrometer, μm
MICROPHOTOGRAPHY	Microphotographie	Mikrokopie
MICROPHOTOMETER	Microphotomètre	Mikrophotometer
MICROPHOTOMETRY	Microphotométrie	Mikrophotometrie
MICROPROJECTOR	Microscope de projection	Mikroprojektor
MICROSCOPE	Microscope	Mikroskop
MICROSCOPE, BINOCULAR	Microcope binoculaire; Loupe binoculaire	Binokular-Mikroskop
MICROSCOPE, COMPARISON	Microscope de comparaison	Vergleichsmikroskop
MICROSCOPE, COMPOUND	Microscope	zusammengesetztes Mikroskop
MICROSCOPE, EXIT PUPIL OF	Pupille de sortie du microscope	Austrittspupille des Mikroskops
MICROSCOPE, FIELD	Microscope de voyage	Reisemikroskop
MICROSCOPE, FLUORESCENCE	Microscope pour fluorescence	Fluoreszenz-Mikroskop
MICROSCOPE, FLYING SPOT	Microscope à flying spot	Flying Spot-Mikroskop
MICROSCOPE, GREENOUGH	Microscope de Greenough	Greenough-Mikroskop
MICROSCOPE, INTERFERENCE	Microscope interférentiel	Interferenzmikroskop
MICROSCOPE, INVERTED	Microscope inversé	umgekehrtes Mikroskop
MICROSCOPE, LIGHT	Microscope optique	Lichtmikroskop
MICROSCOPE, MEASURING	Microscope de mesures	Meßmikroskop
MICROSCOPE, MONOCULAR	Microscope avec tube monoculaire	monokulares Mikroskop
MICROSCOPE, ORE	Microscope de pétrographe	Erzmikroskop
MICROSCOPE, POLARIZED-LIGHT	Microscope polarisant	Polarisationsmikroskop
MICROSCOPE, PORTABLE	Microcope de voyage	Reisemikroskop
MICROSCOPE, PROJECTION	Microscope de projection	Mikroprojektor
MICROSCOPE, REFLECTED LIGHT	Microscope pour lumière réfléchie	Auflichtmikroskop
MICROSCOPE, REFLECTING	Microscope à miroirs	Spiegelmikroskop
MICROSCOPE, SCANNING OPTICAL	Microscope optique à balayage	optisches Scanning-Mikroskop
MICROSCOPE, SIMPLE	Microscope simple	einfaches (einstufiges) Mikroskop
MICROSCOPE, STEREO	Microscope stéréoscopique	Stereomikroskop

MICROSCOPE, TELEVISION	Vidéo-microscope; Télé-microscope	Fernsehmikroskop
MICROSCOPE, TRANSMITTED LIGHT	Microscope pour lumière transmise	Durchlichtmikroskop
MICROSCOPE, TRAVELLING	Microscope de mesures	Meßschlitten-Mikroskop
MICROSCOPE, VIDEO-ENHANCED CONTRAST	Microscope à contraste amplifié par vidéo	Bildkontrastverstärkungs-Mikroskop
MICROSCOPE BASE	Pied du microscope	Mikroskopfuß
MICROSCOPE FLUORIMETRY	Microfluorimétrie	Mikroskop-Fluorometrie
MICROSCOPE INTERFEROMETRY	Micro-Interférométrie	Mikroskop-Interferometrie
MICROSCOPE LAMP	Lampe de microscope	Mikroskopierleuchte
MICROSCOPE LIMB	Potence du microscope	Tubusträger; Stativbügel
MICROSCOPE MIRROR	Miroir d'éclairage du microscope	Mikroskopspiegel
MICROSCOPE PHOTOMETER	Microphotomètre	Mikroskop-Photometer
MICROSCOPE PHOTOMETRY	Microphotometrie	Mikroskop-Photometrie
MICROSCOPE SLIDE	Lame porte-objet	Objektträger
MICROSCOPE STAGE	Platine du microscope	Objekttisch (des Mikroskops)
MICROSCOPE STAND	Statif du microscope	Mikroskopstativ
MICROSCOPE TUBE	Tube du microscope	Mikroskoptubus
MICROSCOPIC	Microscopique	mikroskopisch (mikroskopisch klein)
MICROSCOPICAL	Microscopique	mikroskopisch (im Zusammenhang mit dem Mikroskop)
MICROSCOPY	Microscopie	Mikroskopie
MICROSCOPY, BRIGHT-FIELD	Microscopie en fond clair	Hellfeldmikroskopie
MICROSCOPY, DARKGROUND (DARKFIELD)	Microscopy en fond noir	Dunkelfeldmikroskopie
MICROSCOPY, DISPERSION STAINING	Micro-analyse par dispersion chromatique	Dispersionsfärbungs-Mikroskopie
MICROSCOPY, EPI-	Microscopie en lumière incidente	Auflichtmikroskopie
MICROSCOPY, FLUORESCENCE	Microscopie par fluorescence	Fluoreszenzmikroskopie
MICROSCOPY, INFRA-RED	Microscopie en infrarouge	Infrarot-Mikroskopie
MICROSCOPY, INTERFERENCE	Microscopie interférentielle	Interferenzmikroskopie
MICROSCOPY, LIGHT	Microscopie optique	Lichtmikroskopie
MICROSCOPY, ORE	Microscopie appliquée à la pétrographie	Erzmikroskopie
MICROSCOPY, PHASE-CONTRAST	Microscopie en contraste de phase	Phasenkontrast-Mikroskopie
MICROSCOPY, POLARIZED-LIGHT	Microscopie en lumière polarisée	Polarisationsmikroskopie
MICROSCOPY, QUANTITATIVE	Microscopie quantitative	quantitative Mikroskopie
MICROSCOPY, REFLECTED LIGHT	Microscopie en lumière réfléchie	Auflichtmikroskopie
MICROSCOPY, RELIEF-CONTRAST	Microscopie en contrast de relief	Reliefkontrast-Mikroskopie
MICROSCOPY, TRANSMITTED LIGHT	Microscopie en lumière transmise	Durchlichtmikroskopie
MICROSCOPY, ULTRAVIOLET	Microscopie en ultraviolet	Ultraviolett-Mikroskopie
MICROSCOPY, VIDEO-ENHANCED CONTRAST	Microscopie à contraste amplifié par vidéo	Bildkontrastverstärkungs-Mikroskopie
MICROSPECTROPHOTOMETER	Microspectrophotomètre	Mikrospektralphotometer
MICROSPECTROPHOTMETRY	Microspectrophotometrie	Mikrospektralphotometrie
MICROTOME	Microtome	Mikrotom
MINIMUM RESOLVABLE DISTANCE	Limite de résolution	kleinster auflösbarer Abstand
MIRED	Mired	Mired
MIRROR, DICHROIC	Miroir dichroïque	dichroitischer Spiegel
MODAL ANALYSIS		Modalanalyse
MODE		prozentualer Anteil (einer Komponente am Gesamtbestand)
MODULATION CONTRAST	Contraste de modulation	Modulationskontrast
MONOCHROMAT	Monochromat	Monochromat
MONOCHROMATIC ABERRATIONS	Aberrations géométriques	monochromatische Abbildungsfehler

MONOCHROMATIC RADIATION	Rayonnement monochromatique	monochromatische Strahlung
MONOCHROMATISM	Achromatopsie	Monochromasie; Farbenblindheit
MONOCHROMATOR	Monochromateur	Monochromator
MONOCULAR	&Monoculaire	monokular
MONOCULAR MICROSCOPE	Microscope monoculaire	monokulares Mikroskop
MONOCULAR TUBE	Tube monoculaire	Monokulartubus
MOUNT, LENS	Monture de lentille; Barillet	Linsenfassung
MOUNTANT	Milieu de montage	Einschlußmittel
MOUNTING MEDIUM	Milieu de montage	Einschlußmittel
MOVEMENT, BROWNIAN	Mouvement brownien	Brownsche (Molekular-) Bewegung
MULTIPLE-BEAM INTERFERENCE	Interférences à ondes multiples	Vielstrahlinterferenz
MUSCAE VOLITANTES	Mouches volantes	Mouches volantes
MYOPIA	Myopie	Myopie; Kurzsichtigkeit
NARROW-BAND-PASS FILTER	Filtre à bande étroite	Schmalbandfilter
NATURAL LIGHT	Lumière naturelle	natürliches Licht
NEAR-POINT (OF THE EYE)	Punctum proximum	Nahpunkt (des Auges)
NEAR-POINT DISTANCE	Distance du punctum proximum	Nahpunktsabstand
NEAREST DISTANCE OF DISTINCT VISION	Distance minimale de vision distincte	Nahpunktsabstand
NEARSIGHTEDNESS	Myopie	Kurzsichtigkeit; Myopie
NEGATIVE EYEPIECE	Oculaire négatif	
NEGATIVE LENS	Lentille divergente	Negativlinse
NELSONIAN ILLUMINATION	Éclairage de Nelson	
NEUTRAL-DENSITY FILTER	Filtre gris; Densité neutre	Neutralfilter
NEWTON'S SCALE OF COLOURS	Échelle des teintes de Newton	Newtonsche Interferenzfarben
NICOL PRISM	Nicol	Nicolsches Prisma
NIGHT VISION	Vision de nuit; Vision scotopique	Nachtsehen; skotopisches Sehen
NODAL PLANE	Plan nodal	Knotenpunktsebene
NODAL POINTS	Points nodaux	Knotenpunkte
NOMARSKI DIFFERENTIAL INTERFERENCE CONTRAST	Contraste interférentiel différentiel de Nomarski	differentieller Interferenzkontrast nach Nomarski
NON-DISPARATE SPATIAL VISION	Vision monoculaire du relief	nicht querdisparates Tiefensehen
NORMAL	Normale; Perpendiculaire	Senkrechte; Normale
NORMAL POSITION	Position avec axes parallèles	Normalstellung
NOSEPIECE	Porte-objectif	Objektivhalter; Objektivwechsler
NOSEPIECE, CENTRING	Porte-objectif centrable	zentrierbarer Objektivhalter
NOSEPIECE, REVOLVING	Revolver	Objektivrevolver
NUMBER, FIELD OF VIEW	Indice de champ	Sehfeldzahl
NUMERICAL APERTURE	Ouverure numérique	numerische Apertur
OBJECT	Objet	Objekt
OBJECT, AUXILIARY	Lame auxiliaire; Lame retard	Hilfsobjekt
OBJECT, SELF-LUMINOUS	Objet lumineux par lui-même	selbstleuchtendes Objekt; Selbstleuchter
OBJECT DISTANCE	Distance objet	Objektweite
OBJECT FIELD	Champ objet	Objektfeld
OBJECT FINDER	Lame de repérage	Objektfinder
OBJECT MARKER	Marqueur d'objets	Objektmarkierer
OBJECT PLANE	Plan objet	Objektebene
OBJECT-SIDE APERTURE (OF A MICROSCOPE OBJECTIVE)	Ouverture numérique (d'un objectif de microscope)	objektseitige numerische Apertur (eines Mikroskopobjektivs)
OBJECT SPACE	Espace objet	Objektraum
OBJECT TO PRIMARY-IMAGE DISTANCE	Distance entre l'objet et l'image	Objekt-Zwischenbild-Abstand
OBJECTIVE	Objectif	Objektiv
OBJECTIVE, ACHROMATIC	Objectif achromatique	achromatisches Objektiv
OBJECTIVE, APOCHROMATIC	Objectif apochromatique	apochromatisches Objektiv

OBJECTIVE, DRY	Objectif à sec	Trockenobjektiv
OBJECTIVE, FLAT-FIELD	Objectif à champ plan	Planobjektiv
OBJECTIVE, FLUORITE	Objectif à la fluorine	Fluorit-Objektiv
OBJECTIVE, FREE WORKING DISTANCE OF	Distance frontale d'un objectif	freier Arbeitsabstand eines Objektivs
OBJECTIVE, IMAGE DISTANCE CORRECTION OF	Distance (image) de correction d'un objectif	Bildweitenkorrektion des Objektivs
OBJECTIVE, INFINITY-CORRECTED	Objectif corrigé à l'infini	auf unendlich korrigiertes Objektiv
OBJECTIVE, LOCATING FLANGE OF (OR OBJECTIVE SHOULDER)	Épaulement d'un objectif	Anlagefläche des Objektivs
OBJECTIVE, LONG WORKING-DISTANCE	Objectif à grande distance frontale	Objektiv mit großem Arbeitsabstand
OBJECTIVE, PARFOCALIZING DISTANCE OF	Distance d'équilibrage d'un objectif	Abgleichlänge des Objektivs
OBJECTIVE, REFLECTING	Objectif à miroirs	Spiegelobjektiv
OBJECTIVE, SEMI-APOCHROMATIC	Objectif semi-apochromatique	semiapochromatisches Objektiv
OBJECTIVE, SPRING-LOADED	Objectif en monture rétractable	Objektiv mit Federfassung
OBJECTIVE, STRAIN-FREE	Objectif sans tension	spannungsfreies Objektiv
OBJECTIVE, WORKING DISTANCE OF	Distance frontale de l'objectif	freier Arbeitsabstand des Objektivs
OBJECTIVE-LOCATING SURFACE (OF THE NOSEPIECE)	Face d'appui (sur le revolver)	Objektiv-Anlagefläche (am Objektivrevolver)
OBJECTIVE SHOULDER	Épaulement d'un objectif	Objektiv-Anlagefläche
OBJECTIVE TO PRIMARY IMAGE DISTANCE	Distance entre l'épaulement de l'objectiv et l'image	Objektiv-Zwischenbild-Abstand
OBLIQUE EXTINCTION	Extinction oblique	schiefe Auslöschung
OBLIQUE ILLUMINATION	Éclairage oblique	schiefe Beleuchtung
OBSERVATION TUBE	Tube d'observation	Beobachtungstubus
OCULAR	Oculaire	Okular
OCULAR, SLOTTED	Oculaire à tirette de Wright	Wrightsches Okular
OCULAR DOMINACE	Dominance oculaire	Äugigkeit; Führung eines Auges
OIL, IMMERSION	Huile d'immersion	Immersionsöl
OIL IMMERSION LENS	Objectif à immersion	Immersionsobjektiv
OPAL GLASS	Verre opale	Opalglas
OPTIC AXIS	Axe optique de l'oeil	Sehachse
OPTICAL ACTIVITY	Activité optique; Pouvoir rotatoire	optische Aktivität
OPTICAL ANISOTROPY	Anisotropie optique	optische Anisotropie
OPTICAL AXIAL ANGLE	Angle des axes optiques (d'un cristal biaxe)	optischer Achsenwinkel
OPTICAL AXIS	Axe optique	optische Achse
OPTICAL DENSITY	Densité optique	Extinktion
OPTICAL FITTING DIMENSIONS	Cotes optiques et mécaniques de référence du microscope	Anschlußmaße
OPTICAL INDICATRIX	Ellipsoïde des indices	Indikatrix
OPTICAL PATH	Chemin optique	optischer Weg
OPTICAL PATHLENGTH	Chemin optique	optische Weglänge
OPTICAL PATHLENGTH DIFFERENCE	Différence de chemin optique	optische Weglängendifferenz
OPTICAL ROTATION	Rotation du plan de polarisation	optische Drehung
OPTICAL THICKNESS	Épaisseur optique	optische Dicke (Weglänge)
OPTICAL TUBELENGTH	Longueur optique de tube	optische Tubuslänge
OPTICALLY BIAXIAL	Biaxe	optisch zweiachsig
OPTICALLY UNIAXIAL	Uniaxe	optisch einachsig
OPTICS, GEOMETRICAL	Optique géométrique	geometrische Optik
OPTICS, WAVE	Optique ondulatoire	Wellenoptik
ORDINARY RAY	Rayon ordinaire	ordentlicher Strahl
ORE MICROSCOPE	Microscope de pétrographe	Erzmikroskop
ORE MICROSCOPY	Micropétrographie	Erzmikroskopie
ORGAN, VISION	Système visuel	Sehorgan
ORTHOSCOPIC EYEPIECE	Oculaire orthoscopique (de Kellner)	orthoskopisches Okular

ORTHOSCOPIC OBSERVATION	Observation orthoscopique	orthoskopische Beobachtung
OVERCORRECTION	Surcorrection	Überkorrektion
PANCRATIC	Pancratique	pankratisch
PARABOLOID CONDENSOR	Condenseur parabolique	Paraboloidkondensor
PARALLAX	Parallaxe	Parallaxe
PARALLEL POLARS	Polariseurs parallèles	parallele Polare (Polarisatoren)
PARALLEL RAY BUNDLE	Faisceau de rayons parallèles	paralleles Strahlenbündel
PARALLEL RAY PATH	Marche parallèle de rayons	paralleler Strahlengang
PARAXIAL	Paraxial	paraxial; achsennah
PARAXIAL RAY	Rayon paraxial	Paraxialstrahl
PARFOCAL	Équilibré	parfokal; untereinander abgeglichen
PARFOCALIZING DISTANCE (OF THE EYEPIECE)	Distance d'équilibrage de l'oculaire	Abgleichlänge des Okulars
PARFOCALIZING DISTANCE (OF THE OBJECTIVE)	Distance d'équilibrage de l'objectif	Abgleichlänge des Objektivs
PARTIAL COHERENCE	Cohérence partielle	partielle Kohärenz
PATCH STOP	Obturation centrale	Zentralblende
PATH, LIGHT	Trajet optique; Trajet des rayons lumineux	Strahlengang
PATH, OPTICAL	Trajet optique	Strahlengang
PATH, RAY	Marche des rayons	Strahlengang
PATHLENGTH, OPTICAL (OPTICAL DISTANCE)	Chemin optique	optische Weglänge
PATHLENGTH DIFFERENCE, OPTICAL	Différence de chemin optique	optische Weglängendifferenz
PEAK WAVELENGTH	Longueur d'onde du pic	Scheitelpunktswellenlänge
PENCIL, RAY	Pinceau de rayons	Strahlenbüschel
PHASE	Phase	Phase
PHASE-CONTRAST MICROSCOPY	Microscopie en contraste de phase	Phasenkontrastmikroskopie
PHASE DIFFERENCE	Différence de phase	Phasendifferenz
PHASE-PLATE	Lame de phase	Phasenplatte
PHASE SHIFT	Déphasage	Phasenverschiebung
PHASE VELOCITY	Vitesse de phase	Phasengeschwindigkeit
PHENOMENA, ENTOPTIC	Phénomènes entoptiques	entoptische Erscheinungen
PHOSPHORESCENCE	Phosphorecence	Phosphoreszenz
PHOTOLUMINESCENCE	Photoluminescence	Photolumineszenz
PHOTOMACROGRAPHY	Macrophotographie	Makrophotographie
PHOTOMETER	Photomètre	Photometer
PHOTOMETER HEAD	Tête photométrique	Photometerkopf
PHOTOMETER TUBE	Tête photométrique	Photometertubus
PHOTOMETRIC	Photométrique	photometrisch
PHOTOMETRIC CONTRAST	Contraste photométrique	photometrischer Kontrast
PHOTOMETRIC DIAPHRAGM	Diaphragme du champ photométré	photometrische Blende; Photometerblende
PHOTOMETRIC FIELD	Champ photométré	Photometerfeld
PHOTOMETRIC QUANTITY	Grandeur photométrique	photometrische Größe
PHOTOMETRY	Photométrie	Photometrie
PHOTOMETRY, MICROSCOPE	Microphotométrie	Mikroskop-Photometrie
PHOTOMICROGRAPHY	Photomicrographie	Mikrophotographie
PHOTOMICROGRAPHIC FIELD DIAPHRAGM	Diaphragme du champ photographié	mikrophotographische Feldblende
PHOTOMICROGRAPHIC MAGNIFICATION	Grandissement photographique; Échelle	mikrophotographische Vergrößerung
PHOTON	Photon	Photon
PHOTOPIC VISION	Vision photopique	photopisches Sehen; Tagessehen
PHYSIOLOGICAL CONTRAST	Contraste physiologique	physiologischer Kontrast
PINCUSHION DISTORTION	Distortion en coussinet	kissenförmige Verzeichnung
PLAN-OBJECTIVE (OR PLANO-OBJECTIVE)	Objectif à champ plan	Planobjektiv
PLANE, APERTURE	Ouverture	Aperturebene
PLANE, FIELD	Champ	Feldebene

PLANE, FOCAL	Plan focal	Brennebene
PLANE; IMAGE	Plan image	Bildebene
PLANE, IMAGE, PRIMARY	Plan de l'image intermédiare	Zwischenbildebene
PLANE, NODAL	Plan nodal	Knotenpunktebene
PLANE, OBJECT	Plan objet	Objektebene
PLANE, POLARIZATION	Plan de polarisation	Polarisationsebene
PLANE , PRINCIPAL	Plan principal	Hauptebene
PLANE, REFERENCE, FOR OPTICAL FITTING DIMENSIONS	Plan de référence pour les cotes optiques et mécaniques de référence du microscope	Bezugsebene für die optischen Anschlußmaße
PLANE, VIBRATION	Plan de vibration	Schwingungsebene
PLANE, VIBRATION (OF ELECTROMAGNETIC RADIATION)	Plan de vibration (du rayonnement électro-magnétique)	Schwingungsebene der elektromagnetischen Strahlung
PLANE, POLARIZED LIGHT	Lumière polarisée rectiligne	linear polarisiertes Licht
PLANES, CARDINAL	Plans, cardinaux	Kardinalebenen
PLANES, CONJUGATE	Plans conjugués	konjugierte Ebenen
PLATE, FIRST-ORDER RED	Lame onde (teinte sensible) du premier ordre	Rot -1. Ordnung-Plättchen
PLATE, HALF-WAVE	Lame demi-onde	Lambda-halbe-Plättchen
PLATE, QUARTER-WAVE	Lame quart d'onde	Lambda-viertel-Plättchen
PLATE, RETARDATION	Lame auxiliaire; Lame biréfringente	Phasenplatte; Phasenschieber
PLATE, ZONE	Réseau zôné	Zonenplatte
PLEOCHROISM	Pléochroïsme	Pleochroismus
PLEOCHROISM, REFLECTION	Pléochroïsme de réflexion	Reflexionspleochroismus
POINT, CONVERGENCE	Point de convergence	Konvergenzpunkt
POINT, FIXATION	Point de fixation	Fixierpunkt
POINT COUNTER	Pointcounter	Point Counter
POINT SOURCE (OF LIGHT)	Source ponctuelle; Point lumineux	Punktlichtquelle
POINTS, CARDINAL	Éléments cardinaux	Kardinalpunkte
POINTS, CONJUGATE	Points conjugués	konjugierte Punkte
POINTS, NODAL	Points nodaux	Knotenpunkte
POINTS, PRINCIPAL	Points principaux	Hauptpunkte
POLAR	Polariseur	Polar (Pl: Polare); (Polarisator)
POLARIZATION, DEGREE OF	Taux de polarisation	Polarisationsgrad
POLARIZATION, DIRECTION (OF ELECTROMAGNETIC RADIATION)	Orientation du plan de polarisation (d'une onde éléctromagnétique)	Polarisationsrichtung (der elektromagnetischen Strahlung)
POLARIZATION FIGURE	Figure de convergence	Polarisationsfigur
POLARIZATION PLANE	Plan de polarisation	Polarisationsebene
POLARIZATION STATE	État de polarisation	Polarisationszustand
POLARIZED LIGHT	Lumière polarisée	polarisiertes Licht
POLARIZED-LIGHT MICROSCOPE	Microscope polarisant	Polarisationsmikroskop
POLARIZED-LIGHT MICROSCOPY	Microscopie polarisante	Polarisationsmikroskopie
POLARIZER	Polariseur	Polarisator
POLARIZING FILTER	Filtre polarisant	Polarisationsfilter
POLARIZING INTERFERENCE	Interférences, par double réfraction, en lumière polarisée	Polarisationsinterferenz
POLARIZING PRISM	Polariseur; Prisme de... (Nicol, Foucault, Glacebrook, Glan, etc)	Polarisationsprisma
POLAROID®	Polaroïd®	Polaroid®
POLAROID® FILM	Film Polaroïd®	Polaroid®-Film
POLARS, CROSSED	Polariseurs croisés	gekreuzte Polare
POLARS, PARALLEL	Polariseurs parallèles	parallele Polare
POLARS, UNCROSSED	Polariseurs décroisés	ungekreuzte Polare
POLISHED SECTION	Échantillon poli; Section polie	polierter Anschliff
PORTABLE MICROSCOPE	Microscope de voyage	Reisemikroskop
POSITION, ADDITION	Position avec axes parallèles	Additionsstellung
POSITION, DIAGONAL	Position diagonale	Diagonalstellung

POSITION, EXTINCTION	Position d'extinction	Auslöschungsstellung
POSITION, NORMAL	Position normale	Normalstellung
POSITION, SUBTRACTION	Position avec axes croisés	Subtraktionsstellung
POSITIVE EYEPIECE	Oculaire positif	
POSITIVE LENS	Lentille convergente	Sammellinse
POWER	Puissance (mécanique), (in W)	(1)Leistungsvermögen; (2)Kraft
POWER, COLLECTING (OF A LENS)	Convergence d'une lentille	Sammelwirkung (einer Linse)
POWER, DISPERSIVE	Réfringence;Pouvoir dispersif	Zerstreuungswirkung (einer Linse)
POWER, MAGNIFYING	1)Grandissement (if the lens gives a real image) 2)Grossissement (if the lens gives a virtual image for visual observation at infinity)	Vergrößerung; Vergrößerungs-vermögen
POWER, REFRACTIVE	Vergence; Puissance (Dioptre units)	Brechkraft
POWER, RESOLVING	Pouvoir résolvant	Auflösungsvermögen
PREPARATION	Préparation	(1)Präparat; (2)Präparation
PRESBYOPIA	Presbytie	Alterssichtigkeit; Presbyopie
PRIMARY DIFFRACTION PATTERN (OR IMAGE)	Spectre de diffraction de Probjet; Spectre des fréquences spatiales	primäre Beugungsfigur; primäres Interferenzbild
PRIMARY IMAGE	Image intermédiaire	reelles Zwischenbild
PRIMARY IMAGE DISTANCE (OF THE OBJECTIVE)	Distance de l'image intermédiaire	Zwischenbildweite (des Objektivs)
PRIMARY IMAGE PLANE	Plan de l'image intermédiaire	Zwischenbildebene
PRIMARY INTERFERENCE IMAGE	Spectre de diffraction de l'objet; Spectre des fréquences spatiales	primäres Interferenzbild
PRIMARY MAGNIFICATION	Grandissement de l'objectif	Abbildungsmaßstab des Objektivs
PRIMARY SOURCE	Source primaire de lumiére	primäre Lichtquelle
PRINCIPAL PLANE	Plan principal	Hauptebene
PRINCIPAL POINT	Point principal	Hauptpunkt
PRINCIPAL RAY	Rayon principal	Hauptstrahl
PRINCIPAL VIBRATION DIRECTIONS	Directions principales de vibration	Hauptschwingungsrichtungen
PRISM	Prisme	Prisma
PRISM, DRAWING	Appareil à dessiner	Zeichenprisma
PRISM, NICOL	Prisme de Nicol	Nicolsches Prisma
PRISM, POLARIZING	Polariseur; Prisme de.. (Nicol,Foucault, Glazebrook, Glan etc)	Polarisationsprisma
PRISM, WOLLASTON	Prisme de Wollaston	Wollaston-Prisma
PROJECTION EYEPIECE	Oculaire de projection	Projektionsokular
PROJECTION HEAD	Tête de projection	Projektionskopf
PROJECTION LENS	Systéme optique de projection	Projektiv
PSEUDOSCOPY	Pseudoscopie	Pseudoskopie
PURITY	Pureté	Farbsättigung
PUPIL	Pupille	Pupille
PUPIL, ENTRANCE	Pupille d'entrée	Eintrittspupille
PUPIL, ENTRANCE, OF THE EYE	Pupille d'entrée de l'oeil	Eintrittspupille des Auges
PUPIL, EXIT	Pupille de sortie	Austrittspupille
PURKINJE SHIFT	Phénomène de Purkinje	Purkinje-Phänomen
PURPLE LINE	Ligne des pourpres	Purpurgerade
QUANTITATIVE MICROSCOPY	Microscopie quantitative	quantitative Mikroskopie
QUANTITATY OF LIGHT	Quantité de lumière	Lichtmenge
QUANTITY, RADIOMETRIC	Grandeur énergétique	radiometrische Größe
QUARTER-WAVE PLATE	Lame quart d'onde	Lambda-viertel-Plättchen
QUARTZ WEDGE	Biseau de Quartz; Coin de quartz	Quarzkeil

APPENDIX II

RADIANCE	Luminance énergétique	Strahldichte
RADIANT ENERGY	Énergie rayonnante	Strahlungsmenge
RADIANT EXITANCE	Exitance énergétique	spezifische Ausstrahlung
RADIANT EXPOSURE	Exposition énergétique	Bestrahlung
RADIANT FLUX	Flux énergétique	Strahlungsfluß
RADIANT INTENSITY	Intensité énergétique	Strahlstärke
RADIATION	Rayonnement	Strahlung
RADIATION, BLACK-BODY	Rayonnement du corps noir	Strahlung eines schwarzen Strahlers
RADIATION, COMPLEX	Rayonnement polychromatique	zusammengesetzte Strahlung
RADIATION; INFRA-RED	Rayonnement infra-rouge	Infrarotstrahlung
RADIATION, MONOCHROMATIC	Rayonnement monochromatique	monochromatische Strahlung
RADIATION, ULTRA-VIOLET	Rayonnement ultra-violet	Ultraviolett-Strahlung
RADIATION, VISIBLE	Rayonnement visible	sichtbare Strahlung
RADIATOR, BLACK BODY	Corps noir; Radiateur intégral	Planckscher Strahler; schwarzer Strahler; Hohlraumstrahler
RADIOMETER	Radiomètre	Radiometer; Strahlungsmesser
RADIOMETRIC	Radiométrique	radiometrisch; strahlungsphysikalisch
RADIOMETRIC QUANTITY	Grandeur énergétique	radiometrische Größe
RADIOMETRY	Radiométrie	Radiometrie; Strahlungsmessung
RAMSDEN DISC	Anneau de Ramsden; Cercle oculaire	Ramsdenscher Kreis
RAMSDEN EYEPIECE	Oculaire de Ramsden	Ramsden-Okular
RASTER	Trame	Raster
RAY	Rayon	Strahl
RAY, AXIAL		Axialstrahl
RAY, CONJUGATE PORTIONS OF		konjugierte Strahlanteile
RAY, EXTRAORDINARY	Rayon extraordinaire	außerordentlicher Strahl
RAY, INCIDENT	Rayon incident	einfallender Strahl
RAY, MARGINAL	Rayon marginal	Randstrahl
RAY, ORDINARY	Rayon ordinaire	ordentlicher Strahl
RAY, PARAXIAL	Rayon paraxial	Paraxialstrahl
RAY, PRINCIPAL	Rayon principal	Hauptstrahl
RAY, BUNDLE	Faisceau de rayons	Strahlenbündel
RAY PATH	Marche des rayons	Strahlengang
RAY PATH, PARALLEL	Marche parallèle de rayons	paralleler Strahlengang
RAY PENCIL	Pinceau de rayons	Strahlenbüschel
RAY SPACE		Strahlenraum
REAL IMAGE	Image réelle	reelles Bild
RED, FIRST ORDER (SENSITIVE TINT)	Pourpre teinte sensible du 1° ordre	Rot 1. Ordnung
REFERENCE DIRECTIONS	Directions de référence	Bezugsrichtungen
REFERENCE PLANE (FOR OPTICAL FITTING DIMENSIONS)	Plan der référence (pour les cotes optiques et mécaniques du microscope)	Bezugsebene (für die optischen Anschlußmaße)
REFERENCE VIEWING DISTANCE	Distance conventionnelle d'observation	Bezugssehweite
REFLECTANCE	Facteur de réflexion	Reflexionsgrad
REFLECTED-LIGHT ILLUMINATION	Éclairage en lumière réfléchie	Auflichtbeleuchtung
REFLECTED-LIGHT ILLUMINATOR	Illuminateur vertical	Auflichtilluminator
REFLECTED-LIGHT MICROSCOPE	Microscope en lumière rèflèchie	Auflichtmikroskop
REFLECTED-LIGHT MICROSCOPY	Microscopie en lumière rèflèchie	Auflichtmikroskopie
REFLECTION	Réflexion	Reflexion
REFLECTION, ANGLE OF	Angle de réflexion	Reflexionswinkel
REFLECTION, DIFFUSE	Réflexion diffuse	diffuse Reflexion
REFLECTION,SPECULAR (REFLECTION REGULAR)	Réflexion spéculaire	spiegelnde Reflexion; (reguläre Reflexion)
REFLECTION, TOTAL INTERNAL	Réflexion totale	Totalreflexion
REFLECTION PLEOCHROISM	Pléochroïsme de réflexion	Reflexionspleochroismus
REFLECTION ROTATION	Rotation du plan de polarisation à la réflexion	Reflexionsdrehung

REFLECTOR	Prisme (Miroir) (Lame) d'éclairage	Reflektor
REFRACTION	Réfraction	Brechung
REFRACTION, DOUBLE	Double réfraction	Doppelbrechung
REFRACTION, OF THE EYE	Refraction de l'oeil	Refraktion des Auges
REFRACTION INCREMENT, SPECIFIC		spezifisches Brechungs-inkrement
REFRACTIVE INDEX	Indice de réfraction	Brechungszahl
REFRACTIVE INDEX, PRINCIPAL	Indice extraordinaire principal	Hauptbrechungszahl
REFRACTIVE POWER	Puissance (d'un système optique)	Brechkraft
REFRACTOMETRY	Réfractométrie	Refraktometrie
REFRACTOMETRY, IMMERSION	Réfractométrie par immersion	Immersionsrefraktometrie
RELATIVE DISPERSION	Pouvoir dispersif	relative Dispersion
RELAY LENS	Véhicule	Übertragungslinse (-system)
RELIEF	Relief	Relief
RELIEF CONTRAST	Contraste de relief	Reliefkontrast
RELIEF-CONTRAST MICROSCOPY	Microscopie à contraste de relief	Reliefkontrast-Mikroskopie
RESOLUTION	Résolution	Auflösung
RESOLVABLE DISTANCE, MINIMUM	Limite de résolution	kleinster auflösbarer Abstand
RESOLVED DISTANCE	Distance résolue	aufgelöster Abstand
RESOLVING POWER	Pouvoir résolvant	Auflösungsvermögen
RESOLVING POWER DIFFRACTION LIMITED	Pouvoir résolvant limité par la diffraction	durch Beugung begrenztes Auflösungsvermögen
RETARDATION	Retard	(Phasen-) Verzögerung
RETARDATION PLATE	Lame auxiliaire; Lame retard	Phasenplatte, Phasenschieber
RETICLE	Réticule	Kreuzgitterplatte
RETINA	Rétine	Netzhaut des Auges; Retina
RETINA, CONES OF	Cônes de la rétine	Zapfen der Netzhaut
RETINA, RODS OF	Batonnets de la rétine	Stäbchen der Netzhaut
RETINAL IMAGE	Image rétinienne	Netzhautbild
REVOLVING NOSEPIECE	Révolver	Objektivrevolver
RHEINBERG ILLUMINATION	Éclairage de Rheinberg	
RING LIGHT	Source circulaire	Ringleuchte
RODS (OF THE RETINA)	Batonnets de la rétine	Stäbchen der Netzhaut
RMS THREAD	Filetage RMS	RMS Objektiv-Standardgewinde
ROTATING STAGE	Platine tournante	drehbarer Objekttisch; Drehtisch
ROTATION, OPTICAL	Polarisation rotatoire	optische Drehung
ROTATION, SPECIFIC	Rotation spécifique	spezifische Drehung
SAMPLE	Échantillon	Probe (zu untersuchende)
SATURATION OF COLOUR	Saturation d'une coleur	Farbsättigung
SCALE BAR	Échelle	Maßstabs-Strich
SCALE, IMAGING	Grandissement	Abbildungsmaßstab
SCANNING MICROSCOPE	Microscope à balayage	Scanning-Mikroskop
SCANNING STAGE	Platine à balayage	Scanning-Tisch
SCOTOPIC VISION	Vision scotopique; Vision de nuit	skotopisches Sehen; Nachtsehen
SCREEN	Écran	Schirm; Scheibe
SCREEN, DIFFUSING	Écran diffusant	Streuscheibe
SCREW MICROMETER	Oculaire à vis micrométrique	Okularschraubenmikrometer
SCREW THREAD, FOR OBJECTIVE	Filetage d'un objectif	Objektiv-Anschraubgewinde
SECONDARY SOURCE	Source secondaire	stellvertretende (sekundäre) Lichtquelle
SELF-LUMINOUS OBJECT	Objet lumineux par lui-même	Selbstleuchter
SEMI-APOCHROMAT	Semi-apochromat	Semiapochromat
SENSITIVE-TINT PLATE	Lame teinte sensible	Rot 1. Ordnung-Plättchen
SHEARING INTERFEROMETER	Interféromètre à dédoublement latéral	Shearing-Interferometer
SHORT-WAVE-PASS FILTER	Filtre passe-bas	Kurzpaßfilter
SIGN OF BIREFRINGENCE	Signe de la biréfringence	Vorzeichen der Doppelbrechung

APPENDIX II

SIMPLE MICROSCOPE	Microscope simple	einfaches (einstufiges) Mikroskop
SINE CONDITION	Condition de sinus	Sinusbedingung
SLIDE	Lame porte-objet	Objektträger
SLIDE, HANGING DROP	Lame pour observation en goutte pendante	Hohlschliffobjektträger
SLIDING STAGE	Platine à glissement	Gleittisch
SLOTTED EYEPIECE	Oculaire à tirette de Wright	Wrightsches Okular
SNELL'S LAW	Loi de Descartes	Snelliussches (Brechungs-) Gesetz
SOCIETY THREAD	Filetage RMS	RMS Objektiv-Standardgewinde
SOLID ANGLE	Angle solide	Raumwinkel
SOURCE	Source	Lichtquelle
SOURCE-FOCUSED ILLUMINATION	Éclairage critique	kritische Beleuchtung
SOURCE, EXTENDED	Source étendue	ausgedehnte Lichtquelle
SOURCE, POINT	Source ponctuelle; Point lumineux	Punktlichtquelle
SOURCE, PRIMARY	Source primaire	primäre Lichtquelle
SOURCE, SECONDARY (SUBSTITUTE)	Source secondaire	stellvertretende (sekundäre) Lichtquelle
SPACE, GAUSSIAN	Espace de Gauss	Gaußscher Raum
SPACE, IMAGE	Espace image	Bildraum
SPACE, OBJECT	Espace objet	Objektraum
SPACE, RAY		Strahlenraum
SPATIAL	Spatial	räumlich; Raum- (als Präfix)
SPATIAL VISION	Vision du relief	räumliches Sehen
SPECIFIC ROTATION	Rotation spécifique	spezifische Drehung (Rotationspolarisation)
SPECIMEN	Échantillon; Objet	Präparat
SPECTRAL	Spectral	spektral
SPECTRAL CONCENTRATION (OR SPECTRAL DENSITY)	Densité spectrale	spektrale Dichte; spektrale Konzentration
SPECTRAL LUMINOSITY CURVE	Courbe d'efficacité lumineuse relative spectrale de l'oeil, V(lamda)	spektrale Hellempfindlich-keitskurve; Vλ-Kurve
SPECTRAL LUMINOUS EFICIENCY (OF THE EYE)	Efficacite lumineuse relative spectrale de l'oeil	spektraler Hellempfindlich-keitsgrad
SPECTROPHOTOMETER	Spectrophotomètre	Spektralphotometer
SPECTROPHOTOMETRY	Spectrophotométrie	Spektralphotometrie
SPECTRUM	Spectre	Spektrum
SPECTRUM LOCUS	Spectrum Locus	Spektralfarbenzug
SPECTRUM, CONTINUOUS	Spectre continu	kontinuierliches Spektrum
SPECTRUM, LINE	Spectre de raies	Linienspektrum
SPECTRUM, SECONDARY	Spectre secondaire	sekundäres Spektrum
SPECTRUM, TERTIARY		tertiäres Spektrum
SPECULAR REFLECTION	Réflexion spéculaire	spiegelnde Reflexion; Spiegelung
STAGE (MICROSCOPE STAGE)	Platine (de microscope)	Objekttisch (des Mikroskops)
STAGE, CENTRING	Platine centrable	zentrierbarer Objekttisch
STAGE, COOLING	Platine réfrigérante	Kühltisch
STAGE, HEATING	Platine chauffante	Heiztisch
STAGE, INTEGRATING	Platine intégrante	Integrationstisch
STAGE, LEVELLING	Platine orientable	Anschlifftisch
STAGE, MECHANICAL	Platine à mouvements croisés; Surplatine	Kreuztisch; Objektführer
STAGE, ROTATING	Platine tournante	drehbarer Objekttisch;
STAGE, SCANNING	Platine à balayage	Scanningtisch
STAGE, SLIDING	Platine à glissement	Gleittisch
STAGE, UNIVERSAL	Platine théodolite	Universaldrehtisch
STAGE CLIP	Valet	Präparatklammer (Tischfeder)
STAGE MICROMETER	Micromètre-objet	Objektmikrometer
STAND (MICROSCOPE STAND)	Statif (de microscope)	Stativ; Mikroskopstativ
STAR TEST	Essai au point lumineux	Sterntest
STERADIAN	Stéradian	Steradiant
STEREOLOGY	Stéréologie	Stereologie

STEREOMICROSCOPE	Microscope stéréoscopique	Stereomikroskop
STEREOSCOPY (STEREOPSIS)	Vision stéréoscopique (stéréopsie)	Stereosehen; Stereopsis
*STILB	*Stilb	*Stilb
STOP	Diaphragme	Blende
STOP, CENTRAL	Obturation centrale	Zentralblende
STRABISMUS	Strabisme	Schielen; Strabismus
STRAIGHT EXTINCTION	Extinction droite	gerade Auslöschung
STRAY LIGHT	Lumière parasite	Streulicht
SUB-MICROSCOPIC	Ultramicroscopique	submikroskopisch
SUBSTAGE	Sous-platine	Beleuchtungsapparat
SUBSTAGE, ABBE	Appareil d'éclairage d'Abbe	Abbescher Beleuchtungs-apparat
SUBSTAGE CONDENSER	Condenseur sur sous-platine	Durchlichtkondensor
SUBTRACTION POSITION	Position avec axes croisés	Subtraktionsstellung
SURROUNDING FIELD	Champ périphérique	Umfeld
SYMMETRICAL EXTINCTION	Extinction symétrique	symmetrische Auslöschung
TELESCOPE, AUXILIARY	Viseur de Bertrand	Einstellfernrohr
TELEVISION HEAD	Tête vidéo	Fernsehkopf
TELEVISION MICROSCOPE	Vidéo-microscope; Télé-microscope	Fernsehmikroskop
TEMPERATURE, COLOUR	Température de couleur	Farbtemperatur
TEMPERATURE, DISTRIBUTION	Température de répartition	Verteilungstemperatur
TERTIARY SPECTRUM		tertiäres Spektrum
TEST OBJECT	Objet-test	Testobjekt
THICKNESS, OPTICAL	Épaisseur optique	optische Weglänge
TISSUE, LENS (LENS PAPER)	Papier à nettoyer les lentilles	Linsenpapier
TOP LENS (OF CONDENSER)	Lentille frontale du condenseur	Frontlinse des Kondensors
TOTAL INTERNAL REFLECTION	Réflexion totale	Totalreflexion
TOTAL MAGNIFICATION	Grossissement total	Gesamtvergrößerung
TRANSILLUMINATION	Éclairage par transmission	Durchlichtbeleuchtung
TRANSFER LENS	Véhicule	Übertragungslinse (-system)
TRANSMISSION	Transmission	Transmission
TRANSMITTANCE	Facteur de transmission	Transmissionsgrad
TRANSMITTANCE, LUMINOUS	Facteur de transmission lumineuse	Helltransmissionsgrad
TRANSMITTED LIGHT	Lumière transmise	Durchlicht
TRANSMITTED-LIGHT MICROSCOPE	Microscope pour lumière transmise	Durchlichtmikroskop
TRANSMITTED-LIGHT MICROSCOPY	Microscopie en lumière transmise	Durchlichtmikroskopie
TRAVELLING MICROSCOPE	Microscope de mesures	Meßschlitten-Mikroskop
TRINOCULAR TUBE	Tube trinoculaire	binokularer Phototubus
TRISTIMULUS COLOUR VALUES	Composantes trichromatiques	Normfarbwerte; trichroma-tische Farbmaßzahlen
TUBE	Tube	Tubus
TUBE, BINOCULAR	Tube binoculaire	Binokulartubus
TUBE, BODY	Corps	Mikroskoptubus
TUBE, CAMERA	Tube photographique	Phototubus
TUBE, INTERMEDIATE	Tube intermédiaire	Zwischentubus
TUBE, MONOCULAR	Tube monoculaire	Monokulartubus
TUBE, PHOTOMETER	Tête photométrique	Photometertubus
TUBE, TRINOCULAR	Tube trinoculaire	binokularer Phototubus
TUBE, VIEWING	Tube porte-oculaire	Beobachtungstubus
TUBE FACTOR	Facteur de tube	Tubusfaktor
TUBE HEAD	Tube droit	Tubuskopf
TUBE-LENS	Lentille de tube	Tubuslinse
TUBELENGTH, MECHANICAL	Longeur mécanique de tube	mechanische Tubuslänge
TUBELENGTH, OPTICAL	Longeur optique de tube	optische Tubuslänge
TUBELENGTH CORRECTION LENS	Lentille de correction de longeur de tube	Tubuslängen-Korrektionslinse
TWILIGHT VISION	Vision crépusculaire	Dämmerungssehen; mesopisches Sehen

APPENDIX II

ULTRAMICROSCOPY	Ultramicroscopie	Ultramikroskopie
ULTRAVIOLET MICROSCOPY	Microscopie en Ultra-Violet	Ultraviolettmikroskopie
ULTRAVIOLET RADIATION	Rayonnement Ultra-Violet	Ultraviolettstrahlung
UNCROSSED POLARS	Polariseurs décroisés	ungekreuzte Polare
UNDERCORRECTION	Sous-correction	Unterkorrektion
UNIAXIAL, OPTICALLY	Uniaxe	optisch einachsig
UNILATERAL OBLIQUE ILLUMINATION	Éclairage oblique	einseitig schiefe Beleuchtung
UNIVERSAL STAGE	Platine universelle	Universaldrehtisch
USEFUL MAGNIFICATION	Grossissement utile	förderliche Vergrößerung
VECTOR, LIGHT	Vecteur lumineux	Lichtvektor
VELOCITY, PHASE	Vitesse de phase	Phasengeschwindigkeit
VERGENCE	Vergence	Vergenz (der Fixierlinien)
VERSION (OF THE EYES)	Version (des yeux)	Version (gleichsinnige Bewegung beider Augen)
VERTEX	Pôle	Linsenscheitel
VERTICAL ILLUMINATION	Éclairage incident	(axiale) Auflichtbeleuchtung
VIBRATION DIRECTION	Direction de vibration	Schwingungsrichtung
VIBRATION DIRECTIONS, PRINCIPAL		Hauptschwingungsrichtungen
VIBRATION PLANE	Plan de vibration	Schwingungsebene
VIDEO-ENHANCED CONTRAST MICROSCOPE	Microscope à contraste amplifié par vidéo	Kontrastverstärkungs-Mikroskop
VIDEO-ENHANCED CONTRAST MICROSCOPY	Microscopie à contraste amplifié par vidéo	Kontrastverstärkungs-Mikroskopie
VIEWING ANGLE	Angle visuel; Diamètre apparent	Blickwinkel; Beobachtungswinkel
VIEWING TUBE	Tube porte-oculaire	Beobachtungstubus
VIRTUAL IMAGE	Image virtuelle	virtuelles Bild
VISIBLE RADIATION	Rayonnement visible	sichtbare Strahlung
VISION, DAYLIGHT (OR PHOTOPIC)	Vision de jour (photopique)	Tagessehen; photopisches Sehen
VISION, MESOPIC	Vision crépusculaire (mésopique)	mesopisches Sehen; Dämmerungssehen
VISION, NIGHT (OR SCOTOPIC)	Vision de nuit (scotopique)	Nachtsehen; skotopisches Sehen
VISION, PHOTOPIC	Vision de jour (photopique)	Tagessehen; photopisches Sehen
VISION, SCOTOPIC	Vision de nuit (scotopique)	skotopisches Sehen; Nachtsehen
VISION, SPATIAL	Vision du relief	räumliches Sehen
VISION, STEREO (STEREOPSIS)	Vision stéréscopique (stéréopsie)	Stereosehen; Stereopsis
VISION, SPATIAL, NON-DISPARATE	Vision monoculaire du relief	nicht querdisparates Tiefensehen
VISION, TWILIGHT (OR MESOPIC)	Vision crépusculaire (mésopique)	Dämmerungssehen; mesopisches Sehen
VISUAL ACUITY	Acuité visuelle	Sehschärfe
VISUAL ANGLE	Angle visuel; Diamètre apparent	Gesichtswinkel
VISUAL AXIS (OF THE EYE)	Axe visuel (de l'oeil)	Sehachse (Gesichtslinie) des Auges
VISUAL FIELD	Champ visuel	Sehfeld; Gesichtsfeld
VISUAL FIELD DIAPHRAGM	Diaphragme de champ de l'oculaire	Sehfeldblende
VISUAL ORGAN	Système visuel	Sehorgan
VITREOUS BODY (OF THE EYE)	Corps vitré	Glaskörper (des Auges)
WARM CHAMBER	Chambre chaud	Heizkammer
WAVE NUMBER	Nombre d'onde	Wellenzahl
WAVE OPTICS	Optique ondulatoire	Wellenoptik
WAVE GROUP	Groupe d'onde	Wellenpaket; (Wellengruppe)
WAVELENGTH	Longeur d'onde	Wellenlänge
WAVELENGTH, CENTRAL	Longeur d'onde central	Zentralwellenlänge

WAVELENGTH, COMPLEMENTARY DOMINANT	Longeur d'onde dominante complémentaire	kompensative Wellenlänge
WAVELENGTH, DOMINANT	Longeur d'onde dominante	farbtongleiche Wellenlämge
WAVELENGTH, PEAK	Longeur d'onde du pic	Scheitelpunktwellenlänge
WAVELENGTH BAND	Bande spectrale; Domaine spectral	Wellenlängenbereich
WAVE TRAIN	Train d'onde	Wellenzug
WEDGE, QUARTZ	Biseau de quartz; Coin de quartz	Quarzkeil
WHITE BODY (OR WHITE STANDARD)	Étalon de blanc	Weißstandard
WHITE LIGHT	Lumière blanche	weißes Licht
WHITE STANDARD	Étalon de blanc	Weißstandard
WIDE-FIELD EYEPIECE	Oculaire à grand champ	Großfeldokular
WIDTH, HALF-HEIGHT (BAND)	Largeur de bande à mi-hauteur	Halbwertsbreite
WOLLASTON PRISM	Prisme de Wollaston	Wollaston-Prisma
WORKING DISTANCE	Distance frontale	Arbeitsabstand
ZONE LENS	Lentille de Fresnel	Zonenlinse; Fresnel-Linse
ZONE PLATE	Réseau zôné	Zonenplatte
ZOOM	Zoom; Objectif pancratique	Zoom; pankratisches System

APPENDIX III

Equivalent terms: French–English–German

French	English	German
Aberration	ABERRATION	Abbildungsfehler
Aberration sphérique	SPHERICAL ABERRATION	sphärische Aberration
Absorption	ABSORPTION	Absorption
Achromat	ACHROMAT	Achromat
Achromatopsie	MONOCHROMATISM	Monochromasie; Farbenblindheit
Accommodation	ACCOMMODATION	Akkommodation
Activité optique; Pouvoir rotatoire	OPTICAL ACTIVITY	optische Aktivität
Acuité visuelle	VISUAL ACUITY	Sehschärfe
Adaptation	ADAPTATION	Adaptation
Adaptation à la lumière	LIGHT ADAPTATION	Helladaptation
Adaptation à l'obscurité	DARK ADAPTATION	Dunkeladaptation
Aigrette; Coma	COMA	Coma
Amétropie	AMETROPIA	Ametropie
Amplificateur de luminances	IMAGE INTENSIFIER	Bildvertärker
Amplitude	AMPLITUDE	Amplitude
Amplitude d'accommodation	ACCOMMODATION RANGE	Akkommodationsbereich
Ampoule	BULB; ENVELOPE	Lampenkolben
Analyse d'image	IMAGE ANALYSIS	Bildanalyse
Analyseur	ANALYSER	Analysator
Angle de convergence	CONVERGENCE ANGLE	Konvergenzwinkel
Angle de divergence	DIVERGENCE ANGLE	Divergenzwinkel
Angle des axes optiques dans un cristal biaxe	OPTIC AXIAL ANGLE	optischer Achsenwinkel
Angle d'extinction	EXTINCTION ANGLE	Auslöschungswinkel
Angle d'incidence	ANGLE OF INCIDENCE	Einfallswinkel
Angle du réflexion	ANGLE OF REFLECTION	Reflexionswinkel
Angle limite	CRITICAL ANGLE	kritischer Winkel
Angle solide	SOLID ANGLE	Raumwinkel
Angle visuel	VIEWING ANGLE	Blickwinkel; Betrachtungswinkel;
Ångstrom	ÅNGTRÖM UNIT	Ångströmeinheit
Anisotropie	ANISOTROPY	Anisotropie
Anisotropie optique	OPTICAL ANISOTROPY	optische Anisotropie
Anneau de Ramsden; cercle oculaire	RAMSDEN DISC	Ramsdenscher Kreis
Anneau oculaire	EYEPOINT	Augenkreis
Apertomètre	APERTOMETER	Apertometer
Aplanétique	APLANATIC	aplanatisch
Apochromat	APOCHROMAT	Apochromat
Apodisation	APODIZATION	Apodisation
Appareil à dessiner	DRAWING APPARATUS	Zeichenapparat
Appareil d'éclairage d'Abbe	ABBE SUBSTAGE	Abbescher Beleuchtungsapparat
Appareil de diffraction d'Abbe	ABBE DIFFRACTION APPARATUS	Abbescher Diffraktionsapparat
Artefact	ARTEFACT	Artefakt
ASA	ASA	ASA
Asphérique	ASPHERICAL	asphärisch

Astigmatisme	ASTIGMATISM	Astigmatismus
Autofluorescence	AUTOFLUORESCENCE; PRIMARY FLUORESCENCE	Eigenfluoreszenz
Autoradiographie	AUTORADIOGRAPHY	Autoradiographie
Axe optique	OPTIC AXIS	optische Achse
Axe optique de l'oeil	OPTIC AXIS (VISUAL AXIS) (OF THE EYE)	Gesichtslinie (Sehachse des Auges)
Axe optique d'un cristal	CRYSTAL OPTICAL AXIS	kristalloptische Achse
Axe visuel (d l'oeil)	VISUAL AXIS (OF THE EYE)	optische Achse (des Auges)
Azimuth	AZIMUTH ANGLE	Azimut (-winkel)
Bague de correction	CORRECTION COLLAR	Korrektionsfassung
Batonets de la rétine	RODS (OF THE REINA)	Stäbchen der Netzhaut
Baume du Canada	CANADA BALSAM	Kanadabalsam
Biaxe	OPTICALLY BIAXIAL	optisch zweiachsig
Binoculaire	BINOCULAR	binokular
Biréfringence	BIREFRINGENCE	Betrag der Doppelbrechung
Biréflectance	BIREFLECTANCE	Doppelreflexionsgrad
Biseau de Quartz; Coin de Quartz	QUARTZ WEDGE	Quarzkeil
Bissectrice	BISECTRIX	Bisektrix
Cadre	MASK	Bildmaske
Candela	CANDELA	Candela
Catadioptrique	CATADIOPTRIC	katadioptrisch
Catoptrique	CATOPTRIC	katoptrisch
Cellule de comptage	COUNTING CHAMBER	Zählkammer
Cellule pour numération sanguine	HAEMOCYTOMETER	Blutkörperchenzählkammer
Cercle de moindre diffusion	CIRCLE OF LEAST CONFUSION	Zerstreuungskreis
Cercle oculaire	EYEPOINT	Augenkreis (des Okulars)
Chambre chaud	WARM CHAMBER	Heizkammer
Chambre clair	CAMERA LUCIDA	Camera lucida
Chambre de compression	COMPRESSARIUM	Kompressarium
Chambre de culture	CULTURE CHAMBER	Kulturkammer
Chambre froid	COOLING STAGE	Kühlkammer
Champ	FIELD	Feld
Champ angulaire (d'un oculaire)	ANGLE OF VIEW (OF EYEPIECE)	Bildwinkel (des Okulars)
Champ d'observation	FIELD OF VIEW	Sehfeld
Champ éclairé	ILLUMINATED FIELD	Leuchtfeld
Champ image	IMAGE FIELD	Bildfeld
Champ (objet ou image)	FIELD PLANE	Feldebene
Champ objet	OBJECT FIELD	Objektfeld
Champ périphérique	SURROUNDING FIELD	Umfeld
Champ photometré	PHOTOMETRIC FIELD	photometrisches Feld
Champ visuel	VISUAL FIELD	Gesichtsfeld (Sehfeld)
Changeur de grossissement	MAGNIFICATION CHANGER	Vergrößerungswechsler
Chemin optique	OPTICAL DISTANCE	optischer Abstand
Chromatique	CHROMATIC	farbig; Farb- (als Präfix)
Chromaticité	CHROMATICITY	Farbart
Chromatisme	CHROMATIC ABERRATION	chromatische Aberration
Chromatisme de grandeur; (latéral)	LATERAL CHROMATIC ABERRATION	chromatische Vergrößerung-differenz
Chromatisme longitudinal; (axial)	AXIAL CHROMATIC ABERRATION	axiale chromatische Aberration
Classe de correction	CORRECTION CLASS	Korrektionsklasse
Code de couleurs des objectifs	COLOUR CODE FOR OBJECTIVES	Farbkode (der Maßstabszahl) der Objektive
Cohérence	COHERENCE	Kohärenz
Cohérence partielle	PARTIAL COHERENCE	partielle Kohärenz
Cohérent	COHERENT	kohärent
Colle	CEMENT	Kitt (Feinkitt)
Collimater	COLLIMATE	kollimieren
Collimateur	COLLIMATOR	Kollimator
Collimation	COLLIMATION	Kollimation

Coma; Aigrette	COMA	Coma
Compensateur	COMPENSATOR	Kompensator
Composantes trichromatiques	TRISTIMULUS COLOUR VALUES	Farbmaßzahlen
Condenseur	CONDENSER	Kondensor
Condenseur à lentille frontale escamotable	SWING-OUT TOP LENS CONDENSER	Kondensor mit ausklappbarer Frontlinse
Condenseur aplanétique et achromatique	ACHROMATIC–APLANATIC CONDENSER	achromatisch–aplanatischer Kondensor
Condenseur cardioïde	CARDIOID CONDENSER	Kardioidkondensor
Condenseur d'Abbe	ABBE CONDENSER	Abbescher Kondensor
Condenseur pancratique	PANCRATIC CONDENSER	pankratischer Kondensor
Condenseur parabolique	PARABOLOID CONDENSER	Paraboloidkondensor
Condenseur pour contraste de phase	PHASE-CONTRAST CONDENSER	Phasenkontrastkondensor
Condenseur pour fond noir	DARKGROUND CONDENSER	Dunkelfeldkondensor
Condition de cohérence	COHERENCE CONDITION	Kohärenzbedingung
Condition de sinus	SINE CONDITION	Sinusbedingung
Cône d'immersion	DIPPING CONE	Immersionsansatz
Cônes de la rétine	RODS (OF THE RETINA)	Stäbchen der Netzhaut
Conjugé	CONJUGATE	konjugiert
Contraste	CONTRAST	Kontrast
Contraste de modulation	MODULATION CONTRAST	Modulationskontrast
Contraste de relief	RELIEF CONTRAST	Reliefkontrast
Contraste interférentiel	INTERFERENCE CONTRAST	Interferenzkontrast
Contraste interférentiel différentiel	DIFFERENTIAL INTERFERENCE CONTRAST	differentieller Interferenzkontrast
Contraste interférentiel différentiel de Nomarski	NOMARSKI DIFFERENTIAL INTERFERENCE CONTRAST	differentieller Interferenzkontrast nach Nomarski
Contraste marginal	MARGINAL CONTRAST	Randkontrast
Contraste photométrique	PHOTOMETRIC CONTRAST	Intensitätskontrast; photometrischer Kontrast
Contraste physiologique	PHYSIOLOGICAL CONTRAST	physiologischer Kontrast
Convergence	CONVERGENCE	Konvergenz
Convergence d'une lentille	COLLECTING POWER OF A LENS	Sammelwirkung einer Linse
Coordonnées trichromatiques	CHROMATICITY CO-ORDINATES	Normfarbwertanteile; Normfarbwertkoordinaten
Cornée	CORNEA	Cornea
Corps noir; Radiateur intégral	BLACK-BODY RADIATOR	Hohlraumstrahler
Corps	BODY TUBE	Tubuskörper
Corps vitré	VITREOUS BODY (OF THE EYE)	Glaskörper (des Auges)
Correction	CORRECTION	Korrektion
Cotes optiques et mécaniques de référence du microscope	OPTICAL FITTING DIMENSIONS (OF THE MICROSCOPE)	optische Anschlußmaße (des Mikroskops)
Colorimétrie	COLORIMETRY	Farbmessung
Couleur	COLOUR	Farbe
Couleur achromatique perçue	ACHROMATIC COLOUR	unbunte Farbe
Couleur chromatique perçue	CHROMATIC COULOUR	bunte Farbe
Couleurs interférentielles	INTERFERENCE COLOUR	Interferenzfarbe
Courbe d'effacité lumineuse relative spectrale de l'oeil, V(lamda)	SPECTRAL LUMINOSITY CURVE	spektrale Hellempfindlichkeitskurve des Auges ($V\lambda$-Kurve)
Courbe de dispersion	DISPERSION CURVE	Dispersionskurve
Courbure de champ image	CURVATURE OF IMAGE FIELD	Bildfeldwölbung
Cristallin	LENS OF THE EYE; CRYSTALLINE LENS	Linse des Auges
Croisée de fils	CROSS WIRES; CROSS HAIRS	Fadenkreuz
Croix noire	EXTINCTION CROSS	Auslöschungskreuz
Crown	CROWN	Kronglas
Définition; Piqué (d'une image)	DEFINITION	Bildschärfe
Degré de cohérence	DEGREE OF COHERENCE	Kohärenzgrad
Densité neutre; Filtre gris	NEUTRAL DENSITY FILTER (OR NEUTRAL FILTER)	Graufilter

Densité spectrale	SPECTRAL CONCENTRATION (OR SPECTRAL DENSITY)	spektrale Dichte (spektrale Konzentration)
Déphasage	PHASE-SHIFT	Phasenverschiebung
Diagramme de chromaticité	CHROMATICITY DIAGRAM	Farbdreieck
Diagramme d'extinction	EXTINCTION CURVE	Auslöschungskurve
Diamètre apparent	VISUAL ANGLE	Gesichtswinkel; Sehwinkel
Diaphragme	DIAPHRAGM	Blende
Diaphragme de champ de l'oculaire	VISUAL FIELD DIAPHRAGM	Sehfeldblende
Diaphragme de Bertrand	BERTRAND DIAPHRAGM	Bertrand-Blende
Diaphragme de lampe	COLLECTOR DIAPHRAGM	Kollektorblende
Diaphragme d'ouverture	APERTURE DIAPHRAGM; (ILLUMINATING APERTURE DIAPHRAGM)	Aperturblende; Beleuchtungs-aperturblende
Diaphragme de champ	FIELD DIAPHRAGM	Feldblende
Diaphragme du champ photographié	PHOTOMICROGRAPHIC DIAPHRAGM	mikrophotographische Bildfeldblende
Diaphragme du condenseur	CONDENSER DIAPHRAGM	Kondensorblende
Diaphragme fixe	FIXED DIAPHRAGM	feste Blende
Diaphragme iris	IRIS DIAPHRAGM	Irisblende
Diaphragme photométrique	PHOTOMETRIC DIAPHRAGM	photometrische Blende
Diaphragme pour fond noir	DARKGROUND STOP	Dunkelfeldblende
Dichroïsme	DICHROISM	Dichroismus
Différence de chemin optique	OPTICAL PATHLENGTH DIFFERENCE	optische Weglängendifferenz
Différence de phase	PHASE DIFFERENCE	Phasenunterschied
Diffraction	DIFFRACTION	Beugung
Diffraction de Fraunhofer	FRAUNHOFER DIFFRACTION	Fraunhofersche Beugung
Diffraction de Fresnel	FRESNEL DIFFRACTION	Fresnelsche Beugung
DIN	DIN	DIN
Dioptrie	DIOPTRE	Dioptrie
Dioptrique	DIOPTRIC	dioptrisch
Direction de vibration	VIBRATION DIRECTION	Schwingungsrichtung
Direction d'extinction	EXTINCTION DIRECTION	Auslöschungsrichtung
Directions de référence	REFERENCE DIRECTION	Bezugsrichtungen
Directions principales de vibration	PRINCIPAL VIBRATION DIRECTIONS	Hauptschwingungsrichtungen
Disparation	DISPARITY	Disparation
Dispersion	DISPERSION	Dispersion
Distance conventionelle	REFERENCE VIEWING DISTANCE	Bezugssehweite
Distance d'accommodation	ACCOMODATION DISTANCE	Akkommodationsentfernung
Distance d'équilibrage de l'objectif	PARFOCALIZING DISTANCE (OF THE OBJECTIVE)	Abgleichlänge des Objektivs
Distance d'équilibrage de l'oculaire	PARFOCALIZING DISTANCE (OF THE EYEPIECE)	Abgleichlänge des Okulars
Distance d'image	IMAGE DISTANCE	Bildweite
Distance du punctum proximum	NEAR POINT DISTANCE (NEAREST DISTANCE OF DISTINCT VISION)	Nahpunktsabstand
Distance du punctum remotum	FAR POINT DISTANCE	Fernpunktsabstand
Distance entre la face d'appui de l'objectif et l'image intermediaire	OBJECTIVE TO PRIMARY (INTER-MEDIATE) IMAGE DISTANCE	Objektiv-Zwischenbild-Abstand
Distance entre l'objet et l'image	OBJECT TO PRIMARY (INTER-MEDIATE) IMAGE DISTANCE	Objekt-Zwischenbild-Abstand
Distance focale	FOCAL LENGTH	Brennweite
Distance frontale	WORKING DISTANCE	Arbeitsabstand
Distance frontale d'un objectif	FREE WORKING DISTANCE OF OBJECTIVE	freier Arbeitsabstand (eines Objektivs)
Distance image de correction	CORRECTION FOR IMAGE DISTANCE	Korrektion auf die Bildweite
Distance minimale de vision distincte	NEAREST DISTANCE OF DISTINCT VISION (NEAR POINT DISTANCE)	Nahpunktsabstand
Distance objet	OBJECT DISTANCE	Objektweite
Distance résolue	RESOLVED DISTANCE	kleinster aufgelöster Abstand
Distortion	DISTORTION	Verzeichnung

Distortion en barillet	BARREL DISTORTION	tonnenförmige Verzeichnung
Distortion en coussinet	PINCUSHION DISTORTION	kissenförmige Verzeichnung
Divergence	DIVERGENCE	Divergenz
Domaine des grossissements utiles	USEFUL RANGE OF MAGNIFICATION	förderlicher Vergrößerungsbereich
Dominance oculaire	OCULAR DOMINANCE	Äugigkeit
Double réfraction	DOUBLE REFRACTION; BIREFRACTION	Doppelbrechung
Doublet	DOUBLET	Duplet
Éblouissement	GLARE	Blendung; Streulicht
Échantillon	SAMPLE	zu untersuchende Probe
Écart d'yeux	INTERPUPILLARY DISTANCE	Pupillenabstand
Écart interpupillaire	INTERPUPILLARY DISTANCE	Pupillenabstand
Échantillon	SPECIMEN; PREPARATION	Präparat
Échantillon poli	POLISHED SECTION	polierter Anschliff
Échelle 1)	LATERAL MAGNIFICATION	Vergrößerungsmaßstab
Échelle 2)	SCALE BAR	Maßstabsstrich
Échelle des teintes de Newton	NEWTON'S SCALE OF COLOURS	Newtonsche Interferenzfarben
Éclairage	ILLUMINATION	Beleuchtung
Éclairage annulaire	ANNULAR ILLUMINATION	ringförmige Beleuchtung
Éclairage au fond noir	DARKGROUND ILLUMINATION	Dunkelfeldbeleuchtung
Éclairage axial	AXIAL ILLUMINATION	axiale Beleuchtung
Éclairage critique	SOURCE FOCUSED (CRITICAL ILLUMINATION)	kritische Beleuchtung
Éclairage de Köhler	KÖHLER ILLUMINATION	Köhlersche Beleuchtung
Éclairage en fond clair	BRIGHT-FIELD ILLUMINATION	Hellfeldbeleuchtung
Éclairage en lumière incidente	INCIDENT ILLUMINATION	Auslichtbeleuchtung; einfallende (auffallende) Beleuchtung
Éclairage en lumière transmise	TRANSMITTED LIGHT ILLUMINATION	Durchlichtbeleuchtung
Éclairage en lumière incidente axiale	EPI-ILLUMINATION	axiale Auflichtbeleuchtung
Éclairage oblique	UNILATERAL OBLIQUE ILLUMINATION	einseitig schiefe Beleuchtung
Éclairement énergétique	IRRADIANCE	Bestrahlungsstärke
Éclairement lumineux (en un point d'une surface)	ILLUMINANCE (AT A POINT OF A SURFACE)	Beleuchtungsstärke (an einem Punkt einer Fläche)
Écran	SCREEN	(Auffang-)Schirm; Scheibe
Écran diffusante	DIFFUSING SCREEN	Streuscheibe
Éfficacité lumineuse relative	LUMINOUS EFFICIENCY	visueller Nutzeffekt
Éfficacité lumineuse relative spectrale de l'oeil	SPECTRAL LUMINOUS EFFICIENCY (OF THE EYE)	spektraler Hellempfindlichkeitsgrad
Éfficacité lumineuse spectrale maximale	MAXIMUM SPECTRAL LUMINOUS EFFICACY (OF THE EYE)	Maximalwert des spektralen photometrischen Strahlungsäquivalents
Effet de biréflectance	BIREFLECTION	Doppelreflexion
Éléments cardinaux	CARDINAL ELEMENTS	Kardinalelemente
Ellipsoïde des indices	INDICATRIX	Indikatrix
Emmission	EMMISSION	Abstrahlung
Énergie rayonnante	RADIANT ENERGY	Strahlungsmenge
Épaulement de l'objectif	LOCATING FLANGE OF OBJECTIVE	Anlagefläche des Objektivs
Épaulement de l'oculaire	LOCATING FLANGE OF EYEPIECE	Anlagefläche des Okulars
Equilibré	PARFOCAL	parfokal (untereinander abgeglichen)
Espace de Gauss	GAUSSIAN SPACE	Gaußscher Raum
Espace image	IMAGE SPACE	Bildraum
Espace objet	OBJECT SPACE	Objektraum
Essai au point lumineux	STAR TEST	Sterntest
Étalon de blanc	WHITE BODY (OR WHITE STANDARD)	Weißstandard
État d'accommodation	ACCOMMODATION STATE	Akkommodationszustand

État d'adaptation	ADAPTATION STATE	Adaptationszustand
État d'adaptation à la lumière	STATE OF LIGHT ADAPTATION	Helladaptationszustand
État de polarisation	POLARIZATION STATE	Polarisationszustand
Excitation	EXCITATION	Anregung
Exitance énergétique (en un point d'une surface)	RADIANT EXITANCE (AT A POINT ON A SURFACE)	spezifische Ausstrahlung (an einem Punkt einer Fläche)
Exitance lumineuse (en un point d'une surface)	LUMINOUS EXITANCE (AT A POINT ON A SURFACE)	spezifische Lichtaus-Strahlung (an einem Punkt einer Fläche)
Exposition énergétique (en un point d'une surface)	RADIANT EXPOSURE (AT A POINT ON A SURFACE)	Bestrahlung (an einem Punkt einer Fläche)
Exposition lumineuse; Lumination	LIGHT EXPOSURE	Belichtung
Extinction	EXTINCTION	Auslöschung
Extinction droite	STRAIGHT EXTINCTION	gerade Auslöschung
Extinction oblique	OBLIQUE EXTINCTION	schiefe Auslöschung
Extinction symétrique	SYMMETRICAL EXTINCTION	symmetrische Auslöschung
Face d'appui	LOCATING SURFACE (OR FLANGE)	Anlagefläche
Face d'appui de l'objectif (sur le revolver)	LOCATING SURFACE FOR OBJECTIVE (OF THE NOSEPIECE)	Anlagefläche für das Objektiv (am Revolver)
Face d'appui de l'oculairs (sur le tube)	LOCATING SURFACE FOR EYEPIECE	Anlagefläche für das Okular
Faces d'appui sur le corps	BODY TUBE LOCATING SURFACES	Anlageflächen am Tubus-körper
Facteur de luminance	LUMINANCE FACTOR	Hellbezugswert
Facteur d'extinction	EXTINCTION FACTOR	Extinktionsfaktor
Faisceau de fibres optiques	FIBRE OPTIC	Faseroptik
Facteur de réflexion	REFLECTANCE	Reflexionsgrad
Facteur de transmission	TRANSMITTANCE	Transmissionsgrad
Facteur de tube	TUBE-FACTOR	Tubusfaktor
Faisceau de rayons	RAY BUNDLE	Strahlenbündel
Faisceau de rayons convergents	CONVERGING RAY BUNDLE	konvergentes Strahlenbündel
Faisceau de rayons divergents	DIVERGING RAY BUNDLE	divergentes Strahlenbündel
Faisceau de rayons parallèles	PARALLEL RAY BUNDLE	paralleles Strahlenbündel
Faisceau lumineux	LIGHT BEAM	Lichtbündel
Faisceau incident	INCIDENT BEAM	einfallendes Licht bündel
Figure d'axes	AXIAL FIGURE, CONOSCOPIC FIGURE	Achsenbild; konoskopisches Bild
Figure de convergence	POLARIZATION FIGURE	Polarisationsfigur (-bild)
Figure de diffraction	DIFFRACTION PATTERN	Beugungsfigur
Filament	FILAMENT	Glühwendel
Film Polaroid®	POLAROID FILM®	Polaroidfilm®
Filetage d'un objectif	SCREW THREAD, FOR OBJECTIVE	Objektiv-Anschraubgewinde
Filetage RMS	RMS THREAD	RMS-Objektiv-Standardgewinde
Filtre	FILTER	Filter; Lichtfilter
Filtre à bande étroite	NARROW-BAND-PASS FILTER (OR NARROW BAND FILTER)	Schmalbandfilter
Filtre à large bande	BROAD-BAND-PASS FILTER	Breitbandfilter; Breitband-paßfilter
Filtre anti-calorique	HEAT (OR HEAT-PROTECTION) FILTER	Wärmeschutzfilter
Filtre gris	NEUTRAL DENSITY FILTER	Graufilter
Filtre coloré	COLOUR FILTER	Farbfilter
Filtre de blocage; Filtre d'arrêt	BARRIER FILTER	Sperrfilter
Filtre de contraste	CONTRAST FILTER	Kontrastfilter
Filtre de conversion de température de couleur	COLOUR-CONVERSION FILTER	(Farb-) Konversionsfilter
Filtre d'excitation	EXCITER FILTER	Erregerfilter
Filtre gris	NEUTRAL DENSITY FILTER	Graufilter

Filtre interférentiel	INTERFERENCE FILTER	Interferenzfilter
Filtre passe-bas	SHORT-WAVE-PASS FILTER	Kurzpaßfilter
Filtre passe haut	LONG-WAVE-PASS FILTER	Langpaßfilter
Filtre polarisant	POLARIZING FILTER	Polarisationsfilter
Fixation	FIXATION	Fixieren
Flash électronique	ELECTRONIC FLASH	Elektronenblitz
Flint	FLINT GLASS	Flintglas
Fluorescence	FLUORESCENCE	Fluoreszenz
Fluorescence induite (secondaire)	SECONDARY FLUORESCENCE	Sekundärfluoreszenz
Fluorine	FLUORITE	Fluorit; Flußspat
Fluorimétrie	FLUORIMETRY	Fluorometrie
Fluorochrome	FLUOROCHROME	Fluorochrom; Fluoreszenz-Farbstoff
Flux énergétique	RADIANT FLUX	Strahlungsfluß
Flux lumineux	LUMINOUS FLUX	Lichtstrom
Fond clair	BRIGHT-FIELD	Hellfeld
Fond noir	DARKGROUND; DARKFIELD	Dunkelfeld
Fovea	FOVEA	Fovea (centralis, des Auges) (Netzhautgrube)
Foyer	FOCUS	Brennpunkt
Goniomètre	GONIOMETER	Goniometer; Winkelmesser
Grandeur énergétique	RADIOMETRIC QUANTITY	radiometrische Größe
Grandeur photomètrique	PHOTOMETRIC QUANTITY	photometrische Größe
Grandissement	MAGNIFYING	Vergrößerung
Grandissement (if the lens gives a virtual image for visual observation at infinity)	MAGNIFYING POWER	Vergrößerungsvermögen
Grandissement de l'objectif	MAGNIFICATION POWER (NUMBER) OF OBJECTIVE; PRIMARY MAGNIFICATION	Maßstabszahl des Objektivs
Grandissement latéral	LATERAL MAGNIFICATION; LINEAR MAGNIFICATION	Abbildungsmaßstab; Linearvergrößerung
Grandissement longitudinal (axial)	AXIAL MAGNIFICATION	Axialvergrößerung
Grandissement photographique	PHOTOMICROGRAPHIC MAGNIFICATION	mikrophotographische Vergrößerung
Grandissement transversal	LATERAL MAGNIFICATION	Lateralvergrößerung
Graticule	GRATICULE	Strichplatte
Graticule oculaire	EYEPIECE-GRATICULE	Okularstrichplatte
Grossissement (if the lens gives a real image)	MAGNIFYING POWER	Vergrößerungsvermögen
Grossissement commercial	ANGULAR MAGNIFICATION	Angularvergrößerung
Grossissement commercial (ou conventionnel) de l'oculaire	MAGNIFYING POWER OF EYEPIECE	Okularvergrößerung
Grossissement total	TOTAL MAGNIFICATION	Gesamtvergrößerung
Grossissement total du microscope	MAGNIFIYING POWER OF THE MICROSCOPE	Gesamtvergrößerung des Mikroskops
Grossissement vide	EMPTY MAGNIFICATION	Leervergrößerung
Groupe d'onde	WAVE GROUP; (WAVE PACKET)	Wellengruppe; Wellenpaket
Goutte pendante	HANGING DROP	hängender Tropfen
Halo	HALO	Halo
Hématimètre	HAEMOCYTOMETER	Blutkörperchenzählkammer
Homal®	HOMAL®	Homal®
Huile d'immersion	IMMERSION OIL	Immersionsöl
Humeur aqueuse	AQUEOUS HUMOR	Kammerwasser
Hypermétropie	HYPERMETROPIA; (FARSIGHTEDNESS)	Hypermetropie (Weitsichtigkeit)
Illuminateur à fibres optiques	FIBRE OPTIC ILLUMINATOR	Faseroptik-Leuchte
Image	IMAGE	Bild

Image aérienne	AERIAL IMAGE	Luftbild
Image en fond noir	DARKGROUND IMAGE	Dunkelfeldbild
Image en fond clair	BRIGHT-FIELD IMAGE	Hellfeldbild
Image réelle	REAL IMAGE	reelles Bild
Image intermédiaire	INTERMEDIATE (PRIMARY) IMAGE	reelles Zwischenbild
Image rétienne	RETINAL IMAGE	Netzhautbild
Image virtuelle	VIRTUAL IMAGE	virtuelles Bild
Imagerie confocale	CONFOCAL IMAGING MODE	konfokale Abbildung
Immersion	IMMERSION	Immersion
Immersion homogène	HOMOGENEOUS IMMERSION	homogene Immersion
Immersion homogène de l'objectif (ou du condenseur)	HOMGENEOUS IMMERSION OF OBJECTIVE (OR CONDENSER)	homogene Immersion des Objektivs (oder Kondensors)
Impression rémanente	AFTER-IMAGE	Nachbild
Incandescence	INCANDESCENCE	Glühen
Incidence	INCIDENCE	Einfall; Auffall
Incohérent	INCOHERENT	inkohärent
Indice de champ	FIELD-OF-VIEW-NUMBER	Sehfeldzahl
Indice de réfraction	REFRACTIVE INDEX	Brechzahl; Brechungszahl
Intensité	INTENSITY	Intensität
Intensité énergétique	RADIANT INTENSITY	Strahlstärke
Intensité lumineues	LUMINOUS INTENSITY	Lichtstärke
Interférence	INTERFERENCE	Interferenz
Interférence à deux ondes	DOUBLE BEAM INTERFERENCE	Zweistrahl-Interferenz
Interférences à ondes multiples	MULTIPLE BEAM INTERFRENCE	Vielstrahlinterferenz
Interférences, par double refraction, en lumière polarisée	POLARIZING INTERFERENCE	Polarisationsinterferenz
Interféromètre à dédoublement latéral	SHEARING INTERFEROMETER	Shearing-Interferometer
Interféromètre à deux foyers	DOUBLE-FOCUS INTERFEROMETER	Doppelfokusinterferometer
Interféromètre différentielle	DIFFERENTIAL INTERFEROMETER	differentielles Interfero-meter
Interférométrie	INTERFEROMETRY	Interferometrie
Interstice	HIATUS	Interstitium
Inversion de pseudo-relief	INVERSION	Inversion
Iris	IRIS	Iris (des Auges)
Irradation	IRRADATION	Bestrahlung
ISO	ISO	ISO
Isogyres	ISOGYRES	Isogyren
Isotropie	ISOTROPY	Isotropie
Lame à teinte sensible du 1° ordre	FIRST-ORDER RED PLATE	Plättchen Rot I.Ordnung
Lame auxiliaire; Lame retard	RETARDATION PLATE	Phasenschieber
Lame creuse	HANGING-DROP SLIDE	Hohlschliff-Objektträger
Lame de phase	PHASE PLATE; RETARDATION PLATE	Phasenplatte
Lame de repérage	OBJECT FINDER	Objektfinder
Lame demi-onde	HALF-WAVE PLATE; ½ λ-Plate	Lambda-halbe-Plättchen
Lame d'onde (teinte sensible) du 1° ordre	FIRST- ORDER RED PLATE	Rot I.Ordnung-Plättchen
Lame porte-objet	SLIDE; MICROSCOPE SLIDE	Objektträger
Lame quart d'onde	QUARTER WAVE PLATE	Lambda-viertel-Plättchen
Lame-test d'Abbe	ABBE TEST PLATE	Abbesche Testplatte
Lamelle couvre objet	COVER GLASS; (COVER SLIP)	Deckglas
Lampe	LAMP	Lampe
Lampe de microscope	MICROSCOPE LAMP	Mikroskopierleuchte
Lampe à décharge	DISCHARGE LAMP	(Gas-) Entladungslampe
Lampe à halogène	HALOGEN LAMP	Halogenlampe
Lampe à incandescence	FILAMENT LAMP	Glühlampe; Glühfadenlampe; Glühbirne
Lampe à vapeur de mercure	MERCURY ARC LAMP	Quecksilberdampflampe

Lampe à xénon	XENON ARC LAMP	Xenonlampe
Laser	LASER	Laser
Lemniscate	LEMNISCATE	Lemniskate
Lentille	LENS	Linse
Lentille asphèrique	ASPHERICAL LENS	asphärische Linse
Lentille collectrice	COLLECTOR	Kollektor
Lentille convergente	POSITIVE (CONVERGING) LENS	Sammel- (Positiv-) Linse
Lentille de Bertrand (Amici-Bertrand)	BERTRAND LENS (AMICI-BERTRAND LENS)	Bertrand-Linse (Amici-Bertrand-Linse)
Lentille de champ	FIELD LENS	Feldlinse
Lentille de correction de longueur de tube	TUBELENGTH CORRECTION LENS	Tubuslängen-Korrektionslinse
Lentille de Fresnel	FRESNEL LENS	Fresnellinse
Lentille de tube	TUBE-LENS	Tubuslinse
Lentille divergente	DIVERGING (NEGATIVE) LENS	Zerstreuungs- (Negativ-) Linse
Lentille épaisse	THICK LENS	dicke Linse
Lentille frontale	FRONT LENS	Frontlinse
Lentille ménisque	MENISCUS LENS	Meniskus; Meniskuslinse
Lentille mince	THIN LENS	dünne Linse
Lentille frontale du condenseur	TOP LENS (OF CONDENSER)	oberste Linse (Frontlinse) des Kondensors
Ligne de Becke	BECKE LINE	Beckesche Linie
Ligne de pourpres	PURPLE LINE	Purpurgerade
Ligne principale de visée	FIXATION LINE	Fixierlinie
Lignes isochromatiques	ISOCHROMATIC CURVES	Isochromaten
Limite de résolution	MINIMUM RESOLVABLE DISTANCE	kleinster auflösbarer Abstand
Liquide d'immersion	IMMERSION LIQUID	Immersionsflüssigkeit
Loi de Beer–Lambert	BEER–LAMBERT LAW	Lambert–Beersches Gesetz
Loi de Descartes	SNELL'S LAW	Snelliussches gesetz
Longueur d'équilibrage	ADJUSTMENT LENGTH	Abgleichlänge
Longueur d'onde	WAVELENGTH	Wellenlänge
Longueur d'onde à mi-hauteur	HALF-PEAK-HEIGHT WAVELENGTH	Halbwertswellenlänge
Longueur d'onde centrale	CENTRAL WAVELENGTH	Zentralwellenlänge
Longueur d'onde de pic	PEAK WAVELENGTH	Scheitelpunktswellenlänge
Longueur d'onde dominante	DOMINANT WAVELENGTH	farbtongleiche Wellenlänge
Longueur d'onde dominante complémentaire	COMPLEMENTARY DOMINANT WAVELENGTH	kompensative Wellenlänge
Longueur mécanique de tube	MECHANICAL TUBELENGTH	mechanische Tubuslänge
Longueur optique de tube	OPTICAL TUBELENGTH	optische Tubuslänge
Loupe	MAGNIFIER; LOUPE	Lupe
Loupe à mise au point	FOCUSING MAGNIFIER	Einstell-Lupe
Lumen	LUMEN	Lumen
Lumière	LIGHT (VISIBLE RADIATION)	Licht (sichtbare Strahlung)
Lumière blanche	WHITE LIGHT	weißes Licht
Lumıère diffractée	DIFFRACTED LIGHT	gebeugtes Licht
Lumière directe	DIRECT LIGHT	direktes Licht
Lumière naturelle	NATURAL LIGHT	natürliches Licht
Lumière parasite	STRAY LIGHT	Streulicht
Lumière polarisée	POLARIZED LIGHT	polarisiertes Licht
Lumière polarisée circulaire	CIRCULARLY POLARIZED LIGHT	zirkular polarisiertes Licht
Lumière polarisée elliptique	ELLIPTICALLY POLARIZED LIGHT	elliptisch polarisiertes Licht
Lumière polarisée rectiligne	PLANE-POLARIZED LIGHT	linear polarisiertes Licht
Lumière tramsmise	TRANSMITTED LIGHT	Durchlicht
Luminance	LUMINANCE	Leuchtdichte
Luminance énergétique	RADIANCE	Strahldichte
Lumination	EXPOSURE	Belichtungsvorgang
Luminescence	LUMINESCENCE	Lumineszenz
Luminosité	LUMINOSITY; BRIGHTNESS	Helligkeit
Lux	LUX	Lux

Macrophotographie	PHOTOMACROGRAPHY	Makrophotographie
Marche de rayons	RAY PATH; LIGHT PATH; PATH, OPTICAL	Strahlengang
Marche parallèle de rayons	PARALLEL RAY PATH	paralleler Strahlengang
Marques de couleur des objectifs	COLOUR MARKING OF OBJECTIVES	Farbkennzeichnung (der speziellen Eigenschaften) der Objektive
Marqueur d'objets	OBJECT MARKER	Objektmarkierer
Micro-analyse par dispersion chromatique	DISPERSION STAINING MICROSCOPY	Dispersionsfärbungs-Mikroskopie
Microdensitomètre	MICRODENSITOMETER	Mikrodensitometer
Microduromètre	MICROHARDNESS TESTER	Mikrohärteprüfer
Microfluoromtrie	MICROSCOPE FLUORIMETRY	Mikroskopfluorometrie
Microinterféromètre	MICROINTERFEROMETER	Mikrointerferometer
Microinterférométrie	MICROSCOPE INTERFEROMETRY	Mikroskop-Interferometrie
Microphotographie	MICROPHOTOGRAPHY	Mikrokopie
Micromanipulateur	MICROMANIPULATOR	Mikromanipulator
Micromètre	MICROMETER; MICROMETER SCREW	Mikrometer; Mikrometer-schraube
Micromètre, μm	MICROMETRE, μm	Mikrometer, μm
Micromètre-objet	STAGE MICROMETER	Objektmikrometer
Microphotomètre	MICROPHOTOMETER	Mikrophotometer
Microphotométrie 1)	MICROPHOTOMETRY	Mikrophotometrie
Microphotométrie 2)	MICROSCOPE PHOTOMETRY	Mikroskop-Photometrie
Microscope	MICROSCOPE	Mikroskop
Microscope à balayage	SCANNING MICROSCOPE	Scanning-Mikroskop
Microscope à contraste amplifié par vidéo	VIDEO-ENHANCED CONTRAST MICROSCOPE	Bildkontrastverstärkungs-Mikroskop
Microscope à flying spot	FLYING SPOT MICROSCOPE	Flying Spot - Mikroskop
Microscope à miroirs	REFLECTING MICROSCOPE	Spiegelmikroskop
Microscope avec tube monoculaire	MONOCULAR MICROSCOPE	monokulares Mikroskop
Microscope (composé)	COMPOUND MICROSCOPE	zusammengesetztes Mikroskop
Microscope binoculaire	BINOCULAR MICROSCOPE	Binokular-Mikroskop
Microscope de dissection	DISSECTING MIRCOSCOP	Präpariermikroskop
Microscope de comparaison	COMPARISON MICROSCOPE	Vergleichsmikroskop
Microscope de Greenough	GREENOUGH MICROSCOPE	Greenough-Mikroskop
Microscope de mesures	MEASURING MICROSCOPE	Meßmikroskop
Microscope de pétrographe	ORE MICROSCOPE	Erzmikroskop
Microscope de projection	MICROPROJECTOR	Mikroprojektor
Microscope de voyage	PORTABLE MICROSCOPE; FIELD MICROSCOPE	Reisemikroskop
Microscope interférentiel	INTERFERENCE MICROSCOPE	Interferenzmikroskop
Microscope inversé	INVERTED MICROSCOPE	umgekehrtes Mikroskop
Microscope monoculaire	MONOCULAR MICROSCOPE	monokulares Mikroskop
Microscope optique	LIGHT MICROSCOPE	Licht mikroskop
Microscope optique à balayage	OPTICAL SCANNING MICROSCOPE	optisches Scanning-Mikroskop
Microscope polarisant	POLARIZED LIGHT MICROSCOPE	Polarisationsmikroskop
Microscope pour fluorescence	FLUORESCENCE MICROSCOPE	Fluoreszenzmikroskop
Microscope pour lumière réfléchie	REFLECTED LIGHT MICROSCOPE	Auflichtmikroskop
Microscope simple	SIMPLE MICROSCOPE	einfaches Mikroskop
Microscope stéréoscopique	STEREOMICROSCOPE	Stereomikroskop
Microscopie	MICROSCOPY	Mikroskopie
Microscopie à contraste amplifié par vidéo	VIDEO ENHANCED CONTRAST MICROSCOPY	Bildkontrastverstärkungs-Mikroskopie
Microscopie à contraste de relief	RELIEF-CONTRAST MICROSCOPY	Reliefkontrast-Mikroskopie
Microscopie appliquée à la pétrographie	ORE MICROSCOPY	Erzmikroskopie
Microscopie en contraste de phase	PHASE-CONTRAST MICROSCOPY	Phasenkontrastmikroskopie
Microscopie en contraste de relief	RELIEF-CONTRAST MICROSCOPY	Reliefkontrast-Mikroskopie

APPENDIX III

Microscopie en fond clair	BRIGHT-FIELD MICROSCOPY	Hellfeldmikroskopie
Microscopie en fond noir	DARKGROUND MICROSCOPY	Dunkelfeldmikroskopie
Microscopie en infrarouge	INFRA-RED MICROSCOPY	Infrarot-Mikroskopie
Microscopie en lumière incidente	EPI-MICROSCOPY	Auflichtmikroskopie
Microscopie en lumière transmise	TRANSMITTED LIGHT MICROSCOPY	Durchlichtmikroskopie
Microscopie en Ultra-Violet	ULTRAVIOLET MICROSCOPY	Ultraviolett-Mikroskopie
Microscopie interférentielle	INTERFERENCE MICROSCOPY	Interferenzmikroskopie
Microscopie (optique)	LIGHT MICROSCOPY	Lichtmikroskopie
Microscopie par fluorescence	FLUORESCENCE MICROSCOPY	Fluorezenzmikroskopie
Microscopie polarisante	POLARIZED LIGHT MICROSCOPY	Polarisationsmikroskopie
Microscopie quantitative	QUANTITATIVE MICROSCOPY	quantitative Mikroskopie
Microscopique, 1)	MICROSCOPICAL	mikroskopisch (im Zusammenhang mit dem Mikroskop)
Microscopique, 2)	MICROSCOPIC	mikroskopisch (mikroskopisch klein)
Microspectrophotomètre	MICROSPECTROPHOTOMETER	Mikrospektralphotometer
Microspectrophotométrie	MICROSPECTROPHOTOMETRY	Mikrospektralphotometrie
Microtome	MICROTOME	Mikrotom
Milieu	MEDIUM	medium
Mired	MIRED	Mired
Milieu de montage	MOUNTING MEDIUM; MOUNTANT	Einschlußmittel
Miroir d'éclairage du microscope	MICROSCOPE MIRROR	Mikroskopspiegel
Miroir de Lieberkühn	LIEBERKÜHN (LIEBERKÜHN REFLECTOR)	Lieberkühnspiegel
Miroir dichroïque	DICHROIC (DICHROMATIC) MIRROR	dichroitischer Spiegel
Monochromat	MONOCHROMAT	Monochromat
Monochromateur	MONOCHROMATOR	Monochromator
Monoculaire	MONOCULAR	monokular
Montage d'un échantillon poli	LEVELLING (OF A POLISHED SECTION)	Ausrichten (eines polierten Anschliffs
Monture de lentille	LENS MOUNT	Linsenfassung
Mouches volantes	*MUSCAE VOLITANTES*	Mouches volantes
Mouvement Brownien	BROWNIAN MOVEMENT	Brownsche (Molekular-) Bewegung
Mouvement de mise au point	FOCUSING MECHANISM	Einstellmechanismus
Mouvement de mise au point du microscope	FOCUSING MECHANISM OF THE MICROSCOPE	Einstellmechanismus (des Mikroskops)
Mouvement lent	FINE ADJUSTMENT	Feintrieb
Mouvement lent (du microscope)	FOCUSING MECHANISM (OF THE MICROSCOPE)	Fokussiereinrichtung (des Mikroskops)
Mouvement rapide	COARSE ADJUSTMENT	Grobtrieb
Myopie	MYOPIA; NEARSIGHTEDNESS	Myopie; Kurzsichtigkeit
Nicol	POLARIZING PRISM (NICOL PRISM)	Nicolsches Prisma
Nombre d'Abbe; Constringence	ABBE NUMBER	Abbesche Zahl
Nombre d'onde	WAVE NUMBER	Wellenzahl
Normale	NORMAL	Normale; Senkrechte
Objectif	OBJECTIVE	Objektiv
Objectif achromatique	ACHROMATIC OBJECTIVE	achromatisches Objektiv
Objectif à champ plan	FLAT FIELD OBJECTIVE	Planobjekiv
Objectif à grande distance frontale	LONG WORKING DISTANCE OBJECTIVE	Objektiv mit großem freiem Arbeitsabstand
Objectif à la fluorine	SEMIAPOCHROMAT; *FLUORITE OBJECTIVE	*Fluoritobjektiv
Objectif à miroirs	REFLECTING OBJECTIVE	Spiegelobjektiv
Objectif apochromatique	APOCHROMATIC OBJECTIVE	apochromatisches Objektiv
Objectif à sec	DRY OBJECTIVE	Trockenobjektiv

Objectif corrigé à l'infini	INFINITY-CORRECTED OBJECTIVE	auf unendlich korrigiertes Objektiv
Objectif en monture rétractable	SPRING-LOADED OBJECTIVE	Objektiv mit Federfassung
Objective sans tension	STRAIN-FREE OBJECTIVE	spannungsfreies Objektiv
Objectif sémi-apochromatique	SEMIAPOCHROMAT	Semiapochromat; *Fluorit-Objektiv
Objet	OBJECT	Objekt
Objet lumineux par lui-même	SELF-LUMINOUS OBJECT	Selbstleuchter; selbstleuchtendes Objekt
Objet-test	TEST OBJECT	Testobjekt
Observation conoscopique (ou Conoscopie)	CONOSCOPIC OBSERVATION (OR CONOSCOPY)	konoskopische Beobachtung; Konoskopie
Observation orthoscopique	ORTHOSCOPIC OBSERVATION	orthoskopische Beobachtung
Obturation central	CENTRAL STOP	Zentralblende
Oculaire à aiguille	POINTER EYEPIECE	Zeigerokular
Oculaire à champ plan	FLAT-FIELD EYEPIECE	Planokular
Oculaire à dédoublement d'images	IMAGE SHEARING EYEPIECE	Doppelbildokular
Oculaire à grand champ	WIDE-FIELD EYEPIECE	Großfeldokular
Oculaire à graticule	GRATICULE EYEPIECE	Strichplattenokular
Oculaire à mise au point	FOCUSABLE EYEPIECE	einstellbares Okular
Oculaire à tirette de Wright	SLOTTED EYEPIECE	Wrightsches Oklular
Oculaire à vis micrométrique	MICROMETER SCREW EYEPIECE	Okularschraubenmikrometer
Oculaire compensateur	COMPENSATING EYEPIECE	Kompensationsokular; Kompensokular
Oculaire de comptage	INTEGRATING EYEPIECE	Integrationsokular
Oculaire de Huygens	HUYGENIAN EYEPIECE	Huygenssches Okular
Oculaire de Kellner	KELLNER EYEPIECE	Kellnersches Okular
Oculaire de projection	PROJECTION EYEPIECE	Projektionsokular
Oculaire de Ramsden	RAMSDEN EYEPIECE	Ramsdensches Okular
Oculaire goniométrique	GONIOMETER EYEPIECE	Goniometerokular
Oculaire micrométrique	MICROMETER EYEPIECE	Meßokular
Oculaire négatif	INTERNAL-DIAPHRAGM EYEPIECE	Okular mit Zwischenblende
Oculaire orthoscopique (de Kellner)	ORTHOSCOPIC EYEPIECE	orthoskopisches Okular
Oculaire positif	EXTERNAL-DIAPHRAGM EYEPIECE	Okular mit Vorderblende
Oculaire pour comptage	COUNTING EYEPIECE	Zählokular
Oculaire pour porteurs de lunettes	HIGH POINT EYEPIECE	Brillenträgerokular
Oeil	EYE	Auge
Optique à immersion	IMMERSION LENS	Immersionssystem
Optique de tube	INTERMEDIATE LENS	Zwischenlinse
Optique géométrique	GEOMETRICAL OPTICS	geometrische Optik
Optique ondulatoire	WAVE OPTICS	Wellenoptik
Orientation du plan de polarisation (d'un onde éléctromagnétique	POLARIZATION DIRECTION (OF ELECTROMAGNETIC RADIATION)	Polarisationsrichtung (der elektromagnetischen Strahlung)
Orienté à 45 degrées	DIAGONAL POSITION	Diagonalstellung
Ouverture	APERTURE	Apertur
Ouverture angulaire	ANGULAR APERTURE	Öffnungswinkel
Ouverture numérique	NUMERICAL APERTURE	numerische Apertur
Ouverture numérique dans l'espace image (d'un objectif de microscope)	IMAGE-SIDE NUMERICAL APERTURE (OF A MICROSCOPE OBJECTIVE)	bildseitige numerische Apertur (eines Mikroskopobjektivs)
Ouverture numérique (dans l'espace object d'un objectif de microscope)	OBJECT-SIDE NUMERICAL APERTURE (OF A MICROSCOPE OBJECTIVE)	objektseitige numerische Apertur (eines Mikroskopobjektivs)
Pancratique	PANCRATIC	pankratisch
Papier à nettoyer le lentilles	LENS TISSUE (LENS PAPER)	Linsenpapier
Parralaxe	PARALLAX	Parallaxe

APPENDIX III

Paraxial	PARAXIAL	paraxial; achsennah
Perpendiculaire	NORMAL	Senkrechte; Normale
Phase	PHASE; PHASE ANGLE	Phase; Phasenwinkel
Phénomène de Purkinje	PURKINJE SHIFT	Purkinje-Phänomen
Phénomènes entoptique	ENTOPTIC PHENOMENA	entoptische Erscheinungen
Phosphorescence	PHOSPHORESCENCE	Phosphoreszenz
Photoluminescence	PHOTOLUMINESCENCE	Photolumineszenz
Photomètre	PHOTOMETER	Photometer
Photométrie	PHOTOMETRY	Photometrie
Photométrique	PHOTOMETRIC	photometrisch
Photomicrographie 1)	PHOTOMICROGRAPHY	Mikrophotographie
Photomicrographie 2)	MICROGRAPH	mikrophotographische Aufnahme
Photon	PHOTON	Photon
Pied du microscope	MICROSCOPE BASE	Mikroskopfuß
Pinceau de rayons	RAY PENCIL	Strahlenbüschel
Plan de référence (pour les cotes optiques et mécaniques du microscope)	REFERENCE PLANE FOR OPTICAL FITTING DIMENSIONS	Bezugsebene für die optischen Anschlußmaße
Plan de polarisation	POLARIZATION PLANE	Polarisationsebene
Plan de vibration	VIBRATION PLANE	Schwingungsebene
Plan de vibration (du rayonnement électro-magnétique)	VIBRATION PLANE (OF ELECTROMAGNETIC RADIATION)	Schwingungsebene (der elektromagnetischen Strahlung)
Plan d'image intermédiaire	INTERMEDIATE (PRIMARY) IMAGE PLANE	Zwischenbildebene
Plan focal	FOCAL PLANE	Brennebene
Plan focal image	BACK FOCAL PLANE	hintere Brennebene
Plan focal objet	FRONT FOCAL PLANE	vordere Brennebene
Plan image	IMAGE PLANE	Bildebene
Plan nodal	NODAL PLANE	Knotenpunktsebene
Plan objet	OBJECT PLANE	Objektebene
Plan principal	PRINCIPAL PLANE	Hauptebene
Plan pupillaire	APERTURE PLANE	Aperturebene; Pupille
Plans conjugués	CONJUGATE PLANES	konjugierte Ebenen
Platine à balayage	SCANNING STAGE	Scanning-Tisch
Platine à glissement	SLIDING STAGE	Gleittisch
Platine à mouvements croisés	MECHANICAL STAGE	Kreuztisch
Platine centrable	CENTRING STAGE	zentrierbarer Objekttisch
Platine chauffante	HEATING STAGE	Heiztisch
Platine de comptage	INTEGRATING STAGE	Integrationstisch
Platine (de microscope)	STAGE; MICROSCOPE STAGE	Objekttisch (des Mikroskops)
Platine orientable	LEVELLING STAGE	Anschlifftisch
Platine réfrigérante	COOLING STAGE	Kühltisch
Platine tournante	ROTATING STAGE	Drehtisch
Platine universelle	UNIVERSAL STAGE	Universaldrehtisch
Pléochroïsme	PLEOCHROISM	Pleochroismus
Pleochroïsme de réflexion	REFLECTION PLEOCHROISM	Reflexionspleochroismus
Point de convergence	CONVERGING POINT	Konvergenzpunkt
Point de fixation	FIXATION POINT	Fixierpunkt
Pointcounter	POINT COUNTER	Point Counter
Points conjugués	CONJUGATE POINTS	konjugierte Punkte
Points principaux	PRINCIPAL POINTS	Hauptpunkte
Points nodeaux	NODAL POINTS	Knotenpunkte
Polariseur	POLAR; POLARIZER	Polar; Polarisator
Polariseurs croisés	CROSSED POLARS	gekreuzte Polare (Polarisatoren)
Polariseurs décroisés	UNCROSSED POLARS	ungekreuzte Polare (Polarisatoren)
Polariseurs parallèles	PARALLEL POLARS	parallele Polare (Polarisatoren)
Polarisation rotatoire	OPTICAL ROTATION	optische Drehung
Polaroid®	POLAROID®	Polaroid®
Pôle	VERTEX	Linsenscheitel
Porte-objectif	NOSEPIECE	Objektivhalter (-wechsler)

Porte-objectif centrable	CENTRING NOSEPIECE	zentrierbarer Objektiv-halter (-wechsler)
Posemètre	EXPOSURE METER	Belichtungsmesser
Position avec axes croisés	SUBTRACTION POSITION	Subtraktionsstellung
Position avec axes parallèles	ADDITION POSITION	Additionsstellung
Position d'extinction	EXTINCTION POSITION	Auslöschungsstellung
Position diagonale	DIAGONAL POSITION	Diagonalstellung
Potence du microscope	MICROSCOPE LIMB	Stativbügel; Tubusträger
Pourpre teinte sensible du 1° ordre	FIRST-ORDER RED	Rot I.Ordnung
Pouvoir dispersif	DISPERSIVE POWER; (RELATIVE DISPERSION)	relative Dispersion
Pouvoir résolvant	RESOLVING POWER	Auflösungsvermögen
Pouvoir résolvant limité par la diffraction	DIFFRACTION-LIMITED RESOLVING POWER	beugungsbegrenztes Auflösungsvermögen
Préparation	PREPARATION	Präparation; Präparat
Presbytie	PRESBYOPIA	Alterssichtigkeit; Presbyopie
Prisme	PRISM	Prisma
Prisme (Miroir)(Lame) d'éclairage	REFLECTOR	Reflektor
Prisme de Nicol	NICOL PRISM	Nicolsches Prisma
Prisme de . . . (Nicol, Foucault, Glacebrook, Nicol etc)	POLARIZING PRISM OF . . .	Polarisationsprisma nach . . .
Prisme de Wollaston	WOLLASTON PRISM	Wollaston-Prisma
Profondeur de champ	DEPTH OF FIELD (DEPTH OF SHARPNESS IN OBJECT SPACE)	Schärfentiefe (im Objektraum)
Profondeur de champ image	IMAGING DEPTH	Abbildungstiefe
Profondeur de foyer	DEPTH OF FOCUS (DEPTH OF SHARPNESS IN IMAGE SPACE)	Schärfentiefe (im Bildraum)
Pseudoscopie	PSEUDOSCOPY	Pseudoskopie
Punctum remotum	FAR-POINT OF THE EYE	Fernpunkt des Auges
Puissance d'un système optique	REFRACTIVE POWER	Brechkraft
Pupille	PUPIL	Pupille
Pupille d'entrée	ENTRANCE PUPIL	Eintrittspupille
Pupille d'entrée de l'oeil	ENTRANCE PUPIL OF THE EYE	Eintrittspupille des Auges
Pupille de sortie	EXIT PUPIL	Austrittpupille
Pupille de sortie du microscope	EXIT PUPIL OF THE MICROSCOPE	Austrittspupille des Mikroskops
Pupille d'oeil	PUPIL OF THE EYE	Augenpupille
Pureté	PURITY	Farbsättigung (photometr.)
Radiation visible	VISIBLE RADIATION	sichtbare Strahlung
Radiomètre	RADIOMETER	Strahlungsmesser; Radiometer
Radiometrie	RADIOMETRY	Strahlungsmessung
Radiométrique	RADIOMETRIC	radiometrisch
Rayon	RAY	Strahl
Rayon incident	INCIDENT RAY	einfallender Strahl
Rayon marginal	MARGINAL RAY	Randstrahl
Rayon extrordinaire	EXTRAORDINARY RAY	außerordentlicher Strahl
Rayon ordinaire	ORDINARY RAY	ordentlicher Strahl
Rayon paraxial	PARAXIAL RAY	Paraxialstrahl
Rayon principal	PRINCIPAL RAY	Hauptstrahl
Rayonnement	RADIATION	Strahlung
Rayonnement du corps noir	BLACK-BODY RADIATION	schwarze Strahlung; Hohlraumstrahlung
Rayonnement infra-rouge	INFRA-RED RADIATION	Infrarotstrahlung
Rayonnement monochromatique	MONOCHROMATIC RADIATION	monochromatische Strahlung
Rayonnement polychromatique	COMPLEX RADIATION (COMPOUND RADIATION)	Mischstrahlung
Rayonnement visible	VISIBLE RADIATION	sichtbare Strahlung
Rayonnement ultra-violet	ULTRAVIOLET RADIATION	Ultraviolett-Strahlung
Réflexion	REFLECTION	Reflexion

Réflexion diffuse	DIFFUSE REFLECTION	diffuse Reflexion
Réflexion speculaire	SPECULAR REFLECTION (REGULAR REFLECTION)	spiegelnde Reflexion (reguläre Reflexion)
Réflexion totale	TOTAL INTERNAL REFLECTION	Totalreflexion
Réfraction	REFRACTION	Brechung
Réfractométrie	REFRACTOMETRY	Refraktometrie
Réfractométrie par immersion	IMMERSION REFRACTOMETRY	Immersionsrefraktometrie
Relief	RELIEF	Relief
Réseau de diffraction	DIFFRACTION GRATING	Beugungsgitter
Réseau zôné	ZONE PLATE	Zonenplatte
Résolution	RESOLUTION	Auflösung
Résolution limitée par la diffraction	DIFFRACTION LIMIT (OF RESOLVING POWER)	Beugungsbegrenzung des Auflösungsvermögens
Retard	RETARD	Verzögerung (Phasen-)
Rétine	RETINA	Retina
Revolver	REVOLVING NOSEPIECE	Objektivrevolver
Rotation du plan de polarisation à la réflexion	REFLECTION ROTATION	Reflexionsdrehung
Rotation spécifique	SPECIFIC ROTATION	spezifische Drehung
Saturation d'une couleur	SATURATION OF COLOUR	Farbsättigung (physiol.)
Section polie	POLISHED SECTION	poliertes Objekt
Séparateur de faisceaux	BEAM-SPLITTER	Strahlenteiler
Signe de la biréfringence	SIGN OF BIREFRINGENCE	Vorzeichen der Doppelbrechung
Source	SOURCE	Lichtquelle
Source circulaire	RING LIGHT	Ringleuchte
Source étendue	EXTENDED SOURCE	ausgedehnte Lichtquelle
Source primaire	PRIMARY SOURCE	primäre Lichtquelle
Source ponctuelle	POINT SOURCE	punktförmige Lichtquelle
Source secondaire	SECONDARY SOURCE (SUBSTITUTE)	stellvertretende (sekundäre) Lichtquelle
Source primaire de lumière	PRIMARY SOURCE	primäre Lichtquelle
Sous-correction	UNDERCORRECTION	Unterkorrektion
Sous-platine	SUBSTAGE	Beleuchtungsapparat
Spatial	SPATIAL	räumlich; Raum- (als Präfix)
Spectral	SPECTRAL	spektral
Spectre	SPECTRUM	Spektrum
Spectre continu	CONTINUOUS SPECTRUM	kontinuierliches Spektrum
Spectre de diffraction de l'objet	PRIMARY DIFFRACTION PATTERN	primäres Beugungsbild (Interferenzbild)
Spectre de raies	LINE SPECTRUM	Linienspektrum
Spectre des fréquences spatiales	PRIMARY DIFFRACTION PATTERN	primäre Beugungsfigur
Spectre secondaire	SECONDARY SPECTRUM	sekundäres Spektrum
Spectrophotomètre	SPECTROPOTOMETER	Spektralphotometer
Spectrophotométrie	SPECTROPHOTOMETRY	Spektralphotometrie
Spectrum locus	SPECTRUM LOCUS	Spektralfarbenzug
Statif du microscope	STAND (MICROSCOPE STAND)	Mikroskopstativ
Stéradian	STERADIAN	Steradiant
Stéréologie	STEREOLOGY	Stereologie
Stéréopsie	STEREOPSIS	Stereopsis
Stilb	STILB	*Stilb
Strabisme	STRABISMUS	Schielen; Strabismus
Support de filtre	FILTER TRAY	Filterhalter
Surfacique	AREAL	flächig; Flächen- (als Präfix)
Surplatine	MECHANICAL STAGE	Objektführer
Système d'éclairage	ILLUMINATING SYSTEM	Beleuchtungseinrichtung
Système optique de projection	PROJECTION LENS	Projektionssystem; Projektiv
Système visuel	VISUAL ORGAN	Sehorgan
Tache d'Airy	AIRY DISC	Airysches Beugungsscheibchen
Tache de diffraction	DIFFRACTION DISC	Beugungsscheibchen

Taux de polarisation	DEGREE OF POLARISATION	Polarisationsgrad
Tête de projection	PROJECTION HEAD	Projektionskopf
Tête photographique	CAMERA HEAD	Photokopf
Tête photométrique	PHOTOMETER HEAD	Photometerkopf
Tête pour la caméra de télévision	TELEVISION HEAD	Fernsehkopf
Teinte	HUE	Farbton
Télé-microscope	TELEVISION MICROSCOPE	Fernsehmikroskop
Température de couleur	COLOUR TEMPERATURE	Farbtemperatur
Température de répartition	DISTRIBUTION TEMPERATURE	(Farb-) Verteilungs temperatur
Théorie de la formation des images d'après Abbe	ABBE THEORY (OF IMAGE FORMATION)	Abbesche Theorie der Bildentstehung
Train d'onde	WAVE TRAIN	Wellenzug
Traitement de surface	BLOOMING	Beschichtung
Traitement de surface des lentilles	COATING OF LENS SURFACES	Beschichtung (Vergütung) von Linsenoberflächen
Triangle des couleurs	CHROMATICITY DIAGRAM	Farbdreieck
Trame	RASTER	Raster
Transmission	TRANSMISSION	Transmission
Tube	TUBE	Tubus
Tube binoculaire	BINOCULAR TUBE	Binokulartubus
Tube de comparaison	COMPARISON EYEPIECE	Vergleichokular
Tube coulissant	DRAWTUBE	Ausziehtubus
Tube de démonstration	DEMONSTRATION EYEPIECE	Demonstrationsokular
Tube de microscope	BODY TUBE	Mikroskoptubus
Tube droit	TUBE HEAD	Tubuskopf
Tube intermédiaire	INTERMEDIATE TUBE	Zwischentubus
Tube monoculaire	MONOCULAR TUBE	Monokulartubus
Tube photographique	CAMERA TUBE	Phototubus
Tube porte-oculaire	VIEWING TUBE	Beobachtungstubus
Tube trinoculaire	TRINOCULAR TUBE	binokularer Phototubus; (*Trinokulartubus)
Ultramicroscopie	ULTRAMICROSCOPY	Ultramikroskopie
Ultramicroscopique	SUB-MICROSCOPIC	submikroskopisch
Uniaxial	OPTICALLY UNIAXIAL	optisch einachsig
Vecteur de Fresnel	LIGHT VECTOR	Lichtvektor
Vecteur électrique	LIGHT VECTOR	Lichtvektor
Véhicule	RELAY LENS	Übertragungslinse (-system)
Vergence	VERGENCE	Vergenz (der Fixierlinien)
Verre anti-calorique	HEAT-ABSORBING GLASS	Wärmeschutzglas
Verre de mise au point	FOCUSING SCREEN	Einstellscheibe
Verre dépoli	GROUND GLASS	Mattscheibe
Verre d'oeil	EYELENS	Augenlinse (des Okulars)
Verre opale	OPAL GLASS	Opalglas
Vidéo-microscope	TELEVISION MICROSCOPE	Fernsehmikroskop
Viseur de Bertrand	AUXILIARY TELESCOPE	Einstellfernrohr; Hilfsmikroskop
Vision crépusculaire	TWILIGHT VISION	Dämmerungssehen
Vision de jour	DAYLIGHT VISION	Tagessehen
Vision de nuit	NIGHT VISION	Nachtsehen
Vision du relief	SPATIAL VISION	räumliches Sehen
Vision emmétrope	EMMETROPIA	Emmetropie
Vision mésopique	MESOPIC VISION	mesopisches Sehen
Vision monoculaire du relief	NON-DISPARATE SPATIAL VISION	nicht querdisparates Tiefensehen
Vision normale	EMMETROPIA	Normalsichtigkeit
Vision photopique	PHOTOPIC VISION	photopisches Sehen
Vision scotopique	SCOTOPIC VISION	skotopisches Sehen
Vision stéréoscopique	STEREO VISION (STEREOPSIS)	Sterosehen (Stereopsis)
Vitesse de phase	PHASE VELOCITY	Phasengeschwindigkeit
Zoom; Objectif pancratique	ZOOM; PANCRATIC	Zoom (pankratisches System)

APPENDIX IV

Equivalent terms: German–English–French

German	English	French
Abbesche Testplatte	ABBE TEST PLATE	Lame-test d'Abbe
Abbesche Theorie der Bildentstehung	ABBE THEORY (OF IMAGE FORMATION)	Théorie de la formation des images d'après Abbe
Abbesche Zahl	ABBE NUMBER	Nombre d'Abbe; Constringence
Abbescher Beleuchtungs-apparat	ABBE SUBSTAGE	Appareil d'éclairage d'Abbe
Abbescher Diffraktions-apparat	ABBE DIFFRACTION APPARATUS	Appareil de diffraction d'Abbe
Abbescher Kondensor	ABBE CONDENSER	Condenseur d'Abbe
Abbildungsfehler	ABERRATION	Aberration
Abbildungsmaßstab	LATERAL MAGNIFICATION	Grandissement transversal
Abbildungstiefe	IMAGING DEPTH	Profondeur de champ image
Aberration	ABERRATION	Aberration
Aberration, axiale chromatische	CHROMATIC AXIAL ABERRATION	Chromatisme longitudinal (axial)
Aberration, chromatische	CHROMATIC ABERRATION	Chromatisme
Aberration, laterale chromatische	CHROMATIC LATERAL ABERRATION	Chromatisme de grandeur (latéral)
Aberration, sphärische	SPHERICAL ABERRATION	Aberration sphérique
Absorption	ABSORPTION	Absorption
Abgleichlänge	ADJUSTMENT LENGTH	Longueur d'équilibrage
Abgleichlänge des Okulars	PARFOCALIZING DISTANCE (OF THE EYEPIECE)	Distance d'équilibrage de l'oculaire
Abgleichlänge des Objektivs	PARFOCALIZING DISTANCE (OF THE OBJECTIVE)	Distance d'équilibrage de l'objectif
Abstand, aufgelöster	RESOLVED DISTANCE	Distance résolue
Abstand, kleinster auflösbarer	MINIMUM RESOLVABLE DISTANCE	Limite de résolution
Abstand, optischer	OPTICAL DISTANCE	Chemin optique
Abstrahlung	EMMISSION	Emmission
Achromat	ACHROMAT	Achromat
achromatisch–aplanatischer Kondensor	ACHROMATIC–APLANATIC CONDENSER	Condenseur aplanétique et achromatique
Achse, kristalloptische	CRYSTAL OPTICAL AXIS	Axe optique d'un cristal
Achse, optische	OPTIC AXIS	Axe optique
Achse, optische (des Auges)	VISUAL AXIS (OF THE EYE)	Axe visuel (de l'oeil)
Achsenbild	AXIAL FIGURE	Figure d'axes
Achsenwinkel, optischer	OPTIC AXIAL ANGLE	Angle des axes optiques dans un cristal biaxe
Adaptation	ADAPTATION	Adaptation
Adaptationszustand	ADAPTATION STATE	État d'adaptation
Additionsstellung	ADDITION POSITION	Position avec axes parallèles
Airysches Beugungs-scheibchen	AIRY DISC	Tache d'Airy
Ångströmeinheit	ÅNGSTRÖM UNIT	Ångström
Akkommodation	ACCOMMODATION	Accommodation
Akkommodationsbereich	ACCOMMODATION RANGE	Amplitude d'accommodation
Akkommodationsentfernung	ACCOMMODATION DISTANCE	Distance d'accommodation
Akkommodationszustand	ACCOMMODATION STATE	État d'accomodation

Aktivität, optische	OPTICAL ACTIVITY	Activité optique; Pouvoir rotatoire
Altersichtigkeit	PRESBYOPIA	Presbytie
Ametropie	AMETROPIA	Amétropie
Amplitude	AMPLITUDE	Amplitude
Analysator	ANALYSER	Analyseur
Angularvergrößerung	ANGULAR MAGNIFICATION	Grossissement commercial
Anisotropie	ANISOTROPY	Anisotropie
Anisotropie, optische	OPTICAL ANISOTROPY, ANISOTROPISM	Anisotropie optique
Anlagefläche	LOCATING SURFACE (OR FLANGE)	Face d'appui
Anlagefläche des Objektivs	LOCATING FLANGE OF OBJECTIVE	Épaulement de l'objectif
Anlagefläche des Okulars	LOCATING FLANGE OF EYEPIECE	Épaulement de l'oculaire
Anlagefläche für das Objektiv (am Revolver)	LOCATING SURFACE FOR OBJECTIVE (OF THE NOSEPIECE)	Face d'appui de l'objectif (sur le revolver)
Anlagefläche für das Okular	LOCATING SURFACE FOR EYEPIECE	Face d'appui de l'oculairs
Anlageflächen am Tubuskörper	BODY TUBE LOCATING SURFACES	Faces d'appui sur le corps
Anregung	EXCITATION	Excitation
Anschliff, polierter	POLISHED SECTION	Échantillon poli; Section polie
Anschlifftisch	LEVELLING STAGE	Platine orientable
Anschlußmaße, Bezugsebene für die optischen	REFERENCE PLANE FOR OPTICAL FITTING DIMENSIONS	Plan de référence pour définir les cotes optique
Anschlußmaße, optische (des Mikroskops)	OPTICAL FITTING DIMENSION (OF THE MICROSCOPE)	Cotes optiques et mécaniques de référence du microscope
Anteil, prozentualer (einer Komponente am Gesamtbestand)	MODE	
Apertometer	APERTOMETER	Apertomètre
Apertur	APERTURE	Ouverture
Apertur, numerische	NUMERICAL APERTURE	Ouverture numérique
Apertur, numerische bildseitige (eines Mikroskopobjektivs)	IMAGE-SIDE NUMERICAL APERTURE (OF A MICROSCOPE OBJECTIVE)	Ouverture numérique dans l'espace image (d'un objectif de microscope)
Apertur, numerische objektseitige (eines Mikroskopobjektivs)	OBJECT-SIDE NUMERICAL APERTURE (OF A MICROSCOPE OBJECTIVE)	Ouverture numérique (d'un objectif de microscope)
Aperturblende	APERTURE DIAPHRAGM	Diaphragme d'ouverture
Aperturebene	APERTURE PLANE	Plan pupillaire; Pupille
aplanatisch	APLANATIC	Aplanétique
aplanatisch–achromatischer Kondensor	ACHROMATIC–APLANATIC CONDENSER	Condenseur aplanétique et achromatique
aplanatischer Beleuchtungskegel	APLANATIC CONE	
Apochromat	APOCHROMAT	Apochromat
Apodisation	APODIZATION	Apodisation
Arbeitsabstand	WORKING DISTANCE	Distance frontale
Arbeitsabstand, freier (eines Objektivs)	FREE WORKING DISTANCE OF OBJECTIVE	Distance frontale d'un objectif
Artefakt	ARTEFACT	Artefact
ASA	ASA	ASA
asphärisch	ASPHERICAL	asphérique
asphärische Linse	ASPERICAL LENS	Lentille asphérique
Astigmatismus	ASTIGMATISM	Astigmatisme
Auffall	INCIDENCE	Incidence
auffallende Beleuchtung	INCIDENT ILLUMINATION	Éclairage en lumière incidente
Auffallwinkel	ANGLE OF INCIDENCE	Angle d'incidence
aufgelöster Abstand	RESOLVED DISTANCE	Distance résolue
Auflichtbeleuchtung	INCIDENT ILLUMINATION	Éclairage en lumière incidente
Auflichtbeleuchtung, axiale	EPI-ILLUMINATION	Éclairage en lumière incidente axiale
Auflichtmikroskop	REFLECTED LIGHT MICROSCOPE	Microscope pour lumière réfléchie

APPENDIX IV

Auflichtmikroskopie	EPI-MICROSCOPY	Microscopie en lumière incidente
Auflösung	RESOLUTION	Résolution
auflösbarer Abstand, kleinster	MINIMUM RESOLVABLE DISTANCE	Limite de résolution
Auflösungsvermögen	RESOLVING POWER	Pouvoir résolvant
Auflösungsvermögen(s), Begrenzung des, durch Beugung	DIFFRACTION LIMIT (OF RESOLVING POWER)	Résolution limitée par la diffraction
Auge	EYE	Oeil
Augenlinse (des Okulars)	EYELENS	Verre d'oeil
Augenkreis	EYEPOINT	Anneau oculaire
Augenpupille	PUPIL OF THE EYE	Pupille de l'oeil
Äugigkeit	OCULAR DOMINANCE	Dominance oculaire
Auslöschung	EXTINCTION	Extinction
Auslöschung, gerade	STRAIGHT EXTINCTION	Extinction droite
Auslöschung, schiefe	OBLIQUE EXTINCTION	Extinction oblique
Auslöschung, symmetrische	SYMMETRICAL EXTINCTION	Extinction symétrique
Auslöschungskreuz	EXTINCTION CROSS	Croix noire
Auslöschungskurve	EXTINCTION CURVE	Diagramme d'extinction
Auslöschungsrichtung	EXTINCTION DIRECTION	Direction d'extinction
Auslöschungsstellung	EXTINCTION POSITION	Position d'extinction
Auslöschungswinkel	EXTINCTION ANGLE	Angle d'extinction
Ausrichten (eines polierten Anschliffs)	LEVELLING (OF A POLISHED SECTION)	Montage d'un échantillon poli
außerordentlicher Strahl	EXTRAORDINARY RAY	Rayon extraordinaire
Austrittspupille	EXIT PUPIL	Pupille de sortie
Austrittspupille des Mikroskops	EXIT PUPIL OF THE MICROSCOPE	Pupille de sortie du microscope
Ausziehtubus	DRAWTUBE	Tube coulissant
Autoradiographie	AUTORADIOGRAPHY	Autoradiographie
axiale Beleuchtung	AXIAL ILLUMINATION	Éclairage axial
axiale chromatische Aberration	AXIAL CHROMATIC ABERRATION	Chromatisme longitudinal; (axial)
Axialstrahl	AXIAL RAY	
Axialvergrößerung	AXIAL MAGNIFICATION	Grandissement longitudinal; (axial)
Azimutwinkel	AZIMUTH ANGLE	Azimuth
Beckesche Linie	BECKE LINE	Ligne de Becke
Begrenzung des Auflösungs-vermögens durch Beugung	DIFFRACTION LIMIT OF RESOLVING POWER	Résolution limitée par la diffraction
Beleuchtung	ILLUMINATION	Éclairage
auffallende Beleuchtung	INCIDENT ILLUMINATION	Éclairage en lumière incident
Beleuchtung, axiale	AXIAL ILLUMINATION	Éclairage axial
einfallende Beleuchtung	INCIDENT ILLUMINATION	Éclairage en lumière incidente
Beleuchtung, einseitig schiefe	UNILATERAL OBLIQUE ILLUMINATION	Éclairage oblique
Beleuchtung, Köhlersche	KÖHLER ILLUMINATION	Éclairage de Köhler
Beleuchtung, kritische	SOURCE FOCUSED (CRITICAL) ILLUMINATION	Éclairage critique
Beleuchtung. ringförmige	ANNULAR ILLUMINATION	Éclairage annulaire
Beleuchtungsaperturblende	ILLUMINATING APERTURE DIAPHRAGM	Diaphragme d'ouverture
Beleuchtungseinrichtung	ILLUMINATING SYTEM	Système d'éclairage
Beleuchtungskegel, aplanatischer	APLANATIC CONE	
Beleuchtungsstärke (an einem Punkt einer Fläche)	ILLUMINANCE (AT A POINT ON A SURFACE)	Éclairement lumineux (en un point d'une surface)
Belichtung	LIGHT EXPOSURE	Exposition lumineuse; Lumination
Belichtungsmesser	EXPOSURE METER	Posemètre
Belichtungsvorgang	EXPOSURE	Lumination
Beleuchtungsapparat	SUBSTAGE	Sous-platine

Beleuchtungsapparat, Abbescher	ABBE SUBSTAGE	Appareil d'éclairage d'Abbe
Beobachtungstubus	VIEWING TUBE	Tube porte-oculaire
Beobachtungswinkel	VIEWING ANGLE	Angle visuel; Diamètre apparent
Bertrand-Blende	BERTRAND DIAPHRAGM	Diaphragme de Bertrand
Bertrand-Linse	BERTRAND LENS	Lentille de Bertrand
Beschichtung	BLOOMING	Traitement de surface
Beschichtung (Vergütung) von Linsenoberflächen	COATING OF LENS SURFACES	Traitement de surface des lentilles
Bestrahlung	IRRADIATION	Irradation
Bestrahlung (an einem Punkt einer Fläche)	RADIANT EXPOSURE (AT A POINT ON A SURFACE)	Exposition énergétique (en un point d'une surface)
Bestrahlungsstärke	IRRADIANCE	Éclairement énergétique
Betrachtungswinkel	VIEWING ANGLE	Angle visuel; Diamètre apparent
Betrag der Doppelbrechung	BIREFRINGENCE	Biréfringence
Beugung	DIFFRACTION	Diffraction
Beugung, Fraunhofersche	FRAUNHOFER DIFFRACTION	Diffraction de Fraunhofer
Beugung, Fresnelsche	FRESNEL DIFFRACTION	Diffraction de Fresnel
beugungsbegrenztes Auflösungsvermögen	DIFFRACTION-LIMITED RESOLVING POWER	Pouvoir résolvant limité par la diffraction
Beugungsbegrenzung des Auflösungsvermögens	DIFFRACTION LIMIT OF RESOLVING POWER	Résolution limitée par la diffraction
Beugungsbild,primäres	PRIMARY DIFFRACTION PATTERN	Spectre de diffraction de l'objet; Spectre des fréquences spatiales
Beugungsfigur	DIFFRACTION PATTERN	Figure de diffraction
Beugungsfigur, primäre	PRIMARY DIFFRACTION PATTERN	Spectre de diffraction de l'objet; Spectre des fréquences spatiales
Beugungsgitter	DIFFRACTING GRATING	Réseau de diffraction
Beugungsscheibchen	DIFFRACTION DISC	Tache de diffraction
Bewegung, Brownsche (Molekular-)	BROWNIAN MOVEMENT	Mouvement Brownien
Bezugsebene für die optischen Anschlußmaße	REFERENCE PLANE (FOR OPTICAL FITTING DIMENSIONS)	Plan de référence (pour les cotes optiques et mecaniques du microscope)
Bezugsrichtungen	REFERENCE DIRECTIONS	Directions de référence
Bezugssehweite	REFERENCE VIEWING DISTANCE	Distance conventielle d'observation
binokular	BINOCULAR	Binoculaire
Binokular-Mikroskop	BINOCULAR MICROSCOPE	Microscope binoculaire
Binokulartubus	BINOCULAR TUBE; BINOCULAR HEAD	Tube binoculaire
Bild	IMAGE	Image
Bild, konoskopisches	CONOSCOPIC FIGURE	Figure d'axes
Bild, reelles	REAL IMAGE	Image réelle
Bild, virtuelles	VIRTUAL IMAGE	Image virtuelle
Bildanalyse	IMAGE ANALYSIS	Analyse d'image
Bildebene	IMAGE PLANE	Plan image
Bildfeld	IMAGE FIELD	Champ image
Bildfeldblende, mikrophotographische	PHOTOMICROGRAPHIC DIAPHRAGM	Diaphragme du champ photographié
Bildfeldwölbung	CURVATURE OF IMAGE FIELD	Courbure de champ image
Bildkontrastverstäkungs-Mikroskop	VIDEO-ENHANCED CONTRAST MICROSCOPE	Microscope à contraste amplifié par vidéo
Bildkontrastverstärkungs-Mikroskopie	VIDEO-ENHANCED CONTRAST MICROSCOPY	Microscopie à contraste amplifié par vidéo
Bildmaske	MASK	Cadre
Bildraum	IMAGE SPACE	Espace image
Bildschärfe	DEFINITION	Définition; Piqué (d'une image)
bildseitige numerische Apertur (eines Mikroskop-Objektivs)	IMAGE SIDE NUMERICAL APERTURE (OF A MICROSCOPE OBJECTIVE)	Ouverture numérique dans l'espace image (d'un objectif de microscope)

APPENDIX IV

Bildverstärker	IMAGE INTENSIFIER	Amplificateur de luminances
Bildweite	IMAGE DISTANCE	Distance d'image
Bildwinkel (des Okulars)	ANGLE OF VIEW (OF EYEPIECE)	Champ angulaire (d'un oculaire)
Bisektrix	BISECTRIX	Bissectrice
Blende	DIAPHRAGM	Diaphragme
Blende, feste	FIXED DIAPHRAGM	Diaphragme fixe
Blende, photometrische	PHOTOMETRIC DIAPHRAGM	Diaphragme photométrique
Blendung	GLARE	Éblouissement
Blickwinkel	VIEWING ANGLE	Angle visuel
Blutkörperchenzählkammer	HAEMOCYTOMETER	Hématimètre; Cellule pour numération sanguine
Breitbandfilter	BROAD-BAND-PASS FILTER	Filtre à large bande
Breitbandpaßfilter	BROAD-BAND-PASS FILTER	Filtre à large bande
Brechkraft	REFRACTIVE POWER	Puissance d'un système optique
Brechkraft des Auges	REFRACTIVE POWER OF THE EYE	Refraction de l'oeil
Brechung	REFRACTION	Réfraction
Brechungszahl	REFRACTIVE INDEX	Indice de réfraction
Brechungsinkrement, spezifisches	SPECIFIC REFRACTIVE INCREMENT	
Brechzahl	REFRACTIVE INDEX	Indice de Réfraction
Brennebene	FOCAL PLANE	Plan focal
Brennebene, hintere	BACK FOCAL PLANE	Plan focal image
Brennebene, vordere	FRONT FOCAL PLANE	Plan focal objet
Brennpunkt	FOCUS	Foyer
Brennweite	FOCAL LENGTH	Distance focale
Brownsche (Molekular-) Bewegung	BROWNIAN MOVEMENT	Mouvement Brownien
Brillenträgerokular	HIGH-EYEPOINT EYEPIECE	Oculaire pour porteurs de lunettes
bunte Farbe	CHROMATIC COLOUR	Couleur chromatique perçue
Camera lucida	CAMERA LUCIDA	Chambre clair
Candela	CANDELA	Candela
chromatische Aberration	CHROMATIC ABERRATION	Chromatisme
chromatische Vergrößerungsdifferenz	LATERAL CHROMATIC ABERRATION	Chromatisme de grandeur; (latéral)
Cornea	CORNEA	Cornée
Dämmerungssehen; mesopisches Sehen	TWILIGHT VISION; MESOPIC VISION	Vision mésopique
Deckglas	COVER GLASS; (COVER SLIP)	Lamelle couvre objet
Demonstrationsokular	DEMONSTRATION EYEPIECE	Tube de démonstration
Diagonalstellung	DIAGONAL POSITION	Orienté à 45 degrées; Position diagonale
Dichroismus	DICHROISM	Dichroïsme
dichroitischer Spiegel	DICHROIC (DICHROMATIC) MIRROR	Miroir dichroïque
differentielles Interferometer	DIFFERENTIAL INTERFEROMETER	Interféromètre différentielle
differentieller Interferenzkontrast	DIFFERENTIAL INTERFERENCE CONTRAST	Contraste interférentiel différentiel
diffuse Reflexion	DIFFUSE REFLECTION	Réflexion diffuse
DIN	DIN	DIN
Dioptrie	DIOPTRE	Dioptrie
dioptrisch	DIOPTRIC	Dioptrique
direktes Licht	DIRECT LIGHT	Lumière directe
Disparation	DISPARITY	Disparation
Dispersion	DISPERSION	Dispersion
Dispersion, relative	DISPERSIVE POWER; (RELATIVE DISPERSION)	Pouvoir dispersif
Dispersionsfärbungs-Mikroskopie	DISPERSION STAINING MICROSCOPY	Micro-analyse par dispersion chromatique
Dispersionskurve	DISPERSION CURVE	Courbe de dispersion

divergentes Strahlenbündel	DIVERGING RAY BUNDLE	Faisceau de rayons divergents
Divergenz	DIVERGENCE	Divergence
Divergenzwinkel	DIVERGENCE ANGLE	Angle de divergence
Doppelbildokular	IMAGE SHEARING EYEPIECE	Oculaire à dédoublement d'images
Doppelbrechung	DOUBLE REFRACTION; BIREFRACTION	Double réfraction
Doppelbrechung, Betrag der	BIREFRINGENCE	Biréfringence
Doppelfokusinterferometer	DOUBLE-FOCUS INTERFEROMETER	Interféromètre a deux foyers
Doppelreflexion	BIREFLECTION	Effet de biréflectance
Doppelreflexionsgrad	BIREFLECTANCE	Biréflectance
drehbarer Objekttisch	ROTATING STAGE	Platine tournante
Drehung, optische	OPTICAL ROTATION	Polarisation rotatoire
Drehung, spezifische	SPECIFIC ROTATION	Rotation spécifique
Dunkeladaptation	DARK ADAPATION	Adaptation à l'obscurité
Dunkelfeld	DARKGROUND; DARKFIELD	Fond noir
Dunkelfeldbeleuchtung	DARKGROUND ILLUMINATION	Éclairage en fond noir
Dunkelfeldbild	DARKGROUND IMAGE	Image en fond moir
Dunkelfeldblende	DARKGROUND STOP	Diaphragme pour fond noir
Dunkelfeldkondensor	DARKGROUND CONDENSER	Condenseur pour fond noir
Dunkelfeldmikroskopie	DARKGROUND MICROSCOPY	Microscopie en fond noir
Duplet	DOUBLET	Doublet
Durchlicht	TRANSMITTED LIGHT	Lumière transmise
Durchlichtbeleuchtung	TRANSMITTED LIGHT ILLUMINATION	Éclairage en lumière transmise
Durchlichtmikroskopie	TRANSMITTED LIGHT MICROSCOPY	Microscopie en lumière transmise
Ebenen, konjugierte	CONJUGATE PLANES	Plans conjugués
Eigenfluoreszenz	AUTOFLUORESCENCE; FLUORESCENCE PRIMARY	Autofluorescence
einachsig, optisch	OPTICALLY UNIAXIAL	Uniaxe
einfaches Mikroskop	SIMPLE MICROSCOPE	Microscope simple
Einfall	INCIDENCE	Incidence
einfallende Beleuchtung	INCIDENT ILLUMINATION	Éclairage en lumière incidente
einfallendes Licht bündel	INCIDENT BEAM	Faisceau incident
einfallender Strahl	INCIDENT RAY	Rayon incident
Einfallswinkel	ANGLE OF INCIDENCE	Angle d'incidence
Einschlußmittel	MOUNTING MEDIUM; MOUNTANT	Milieu de montage
einseitig schiefe Beleuchtung	UNILATERAL OBLIQUE ILLUMINATION	Éclairage oblique
Einstellfernrohr	AUXILIARY TELESCOPE	Viseur de Bertrand
Einstell-Lupe	FOCUSING MAGNIFIER	Loupe de mise au point
Einstellmechanismus	FOCUSING MECHANISM	Mouvement de mise au point
Einstellmechanismus (des Mikroskops)	FOCUSING MECHANISM OF THE MICROSCOPE	Mouvement de mise au point du microscope
Einstellscheibe	FOCUSING SCREEN	Verre de mise au point
einstufiges Mikroskop	SIMPLE MICROSCOPE	Microscope simple
Eintrittspupille	ENTRANCE PUPIL	Pupille d'entrée
Eintrittspupille des Auges	ENTRANCE PUPIL OF THE EYE	Pupille d'entrée de l'oeil
Elektronenblitz	ELECTRONIC FLASH	Flash électronique
elliptisch polarisiertes Licht	ELLIPTICALLY POLARIZED LIGHT	Lumière polarisée elliptique
Emission; Abstrahlung	EMISSION	Emission
Emmetropie; Normalsichtigkeit	EMMETROPIA	Vision normale; Vision emmétrope
(Gas-) Entladungslampe	DISCHARGE LAMP	Lampe à décharge
entoptische Erscheinungen	ENTOPTIC PHENOMENA	Phénomènes entoptiques
Erregerfilter	EXCITER FILTER	Filtre d'excitation
Erscheinungen, entoptische	ENTOPTIC PHENOMENA	Phénomènes entoptiques
Erzmikroskop	ORE MICROSCOPE	Microscope de pétrographe
Erzmikroskopie	ORE MICROSCOPY	Microscopie appliquée à la pétrographie
Extinktion; Auslöschung	EXTINCTION	Extinction

APPENDIX IV

Extinktionsfaktor	EXTINCTION FACTOR	Facteur d'extinction
Fadenkreuz	CROSS WIRES; CROSS HAIRS	Croisée de fils
Farb- (als präfix)	CHROMATIC	Chromatique
Farbart	CHROMATICITY	Chromaticité
Farbdreieck	CHROMATICITY DIAGRAM	Triangle des couleurs; Diagramme de chromaticité
Farbe	COLOUR	Couleur
Farbe, bunte	CHROMATIC COLOUR	Couleur chromatique perçue
Farbe, unbunte	ACHROMATIC COLOUR	Couleur achromatique perçue
Farbfilter	COLOUR FILTER	Filtre coloré
farbig	CHROMATIC	Chromatique
Farbmaßzahlen	TRISTIMULUS COLOUR VALUES	Composantes trichromatiques
Farbmessung	COLORIMETRY	Colorimétrie
Farbkode (der Maßstabszahl) der Objektive	COLOUR CODE OF OBJECTIVES	Code de couleurs des objectifs
Farbkennzeichnung (der speziellen Eigenschaften) der Objektive	COLOUR MARKING OF OBJECTIVES	Marques de couleur des objectifs
Farbkonversionsfilter	COLOUR-CONVERSION FILTER	Filtre de conversion de température de couleur
Farbort (in der Normfarb-tafel)	SPECTRUM LOCUS	Spectrum locus
Farbsättigung (photometr.)	PURITY	Pureté
Farbsättigung (physiol.)	SATURATION OF COLOUR	Saturation d'une couleur
Farbtemperatur	COLOUR TEMPERATURE	Température de couleur
Farbton	HUE	Teinte
farbtongleiche Wellenlänge	DOMINANT WAVELENGTH	Longueur d'onde dominante
Faseroptik	FIBRE OPTIC	Faisceau de fibres optiques
Faseroptik-Leuchte	FIBRE OPTIC ILLUMINATOR	Illuminateur à fibres optiques
Feintrieb	FINE ADJUSTMENT	Mouvement lent
Feld	FIELD	Champ
Feld, photometrisches	PHOTOMETRIC FIELD	Champ photometré
Feldblende	FIELD DIAPHRAGM	Diaphragme de champ
Feldebene	FIELD PLANE	Champ (objet ou image)
Feldlinse	FIELD-LENS	Lentille de champ
Fernpunkt des Auges	FAR-POINT OF THE EYE	Punctum remotum
Fernpunktsabstand	FAR-POINT DISTANCE	Distance du punctum remotum
Fernsehkopf	TELEVISION HEAD	Tête pour caméra de télévision
Fernsehmikroskop	TELEVISION MICROSCOPE	Vidéo-microscope; Télé-microscope
feste Blende	FIXED DIAPHRAGM	Diaphragme fixe
Filter	FILTER	Filtre
Filterhalter	FILTER TRAY	Support de filtre
Fixieren	FIXATION	Fixation
Fixierlinie	FIXATION LINE	Ligne principale de visée
Fixierpunkt	FIXATION POINT	Point de fixation
Flächen- (als Präfix)	AREAL	Surfacique
flächig	AREAL	Surfacique
Flächenvergrößerung	AREAL MAGNIFICATION	
Fraunhofersche Beugung	FRAUNHOFER DIFFRACTION	Diffraction de Fraunhofer
Fresnelsche Beugung	FRESNEL DIFFRACTION	Diffraction de Fresnel
Flintglas	FLINT GLASS	Flint
Fluoreszenz	FLUORESCENCE	Fluorescence
Fluoreszenzfarbstoff	FLUOROCHROME	Fluorochrome
Fluoreszenzmikroskop	FLUORESCENCE MICROSCOPE	Microscope pour fluorescence
Fluoreszenzmikroskopie	FLUORESCENCE MICROSCOPY	Microscopie par fluorescence
Fluorit	FLUORITE	Fluorine
*Fluoritobjektiv	SEMIAPOCHROMAT; *FLUORITE OBJECTIVE	Objectif à la fluorine
Flußspat	FLUORITE	Fluorine
Fluorometrie	FLUORIMETRY	Fluorimétrie
Fluorochrom	FLUOROCHROME	Fluorochrome
Flying Spot-Mikroskop	FLYING SPOT MICROSCOPE	Microscope à flying spot

Fokussiereinrichtung (des Mikroskops)	FOCUSING MECHANISM (OF THE MICROSCOPE)	Mouvement lent (du microscope)
Photolumineszenz	PHOTOLUMINESCENCE	Photoluminescence
Fovea (centralis, des Auges)	FOVEA	Fovea
Fraunhofersche Beugung	FRAUNHOFER DIFFRACTION	Diffraction des Fraunhofer
freier Arbeitsabstand des Objektivs	FREE WORKING DISTANCE OF OBJECTIVE	Distance frontale d'un objectif
Fresnellinse	FRESNEL LENS	Lentille de Fresnel
Fresnelsche Beugung	FRESNEL DIFFRACTION	Diffraction de Fresnel
Frontlinse	FRONT LENS	Lentille frontale
Führung eines Auges (Äugigkeit)	OCULAR DOMINANCE	Dominance oculaire
Gasentladungslampe	DISCHARGE LAMP	Lampe à décharge
Gaußscher Raum	GAUSSIAN SPACE	Espace de Gauss
gebeugtes Licht	DIFFRACTED LIGHT	Lumière diffractée
geometrische Optik	GEOMETRICAL OPTICS	Optique géométrique
gerade Auslöschung	STRAIGHT EXTINCTION	Extinction droite
Gesamtvergrößerung	TOTAL MAGNIFICATION	Grossissement total
Gesamtvergrößerung des Mikroskops	MAGNIFYING POWER OF THE MICROSCOPE	Grossissement total du microscope
Gesichtsfeld (Sehfeld)	VISUAL FIELD	Champ visuel
Gesichtslinie (Sehachse) des Auges	OPTIC AXIS (VISUAL AXIS) OF THE EYE	Axe optique de l'oeil
Gesichtswinkel	VISUAL ANGLE	Angle visuel; Diamètre apparent
Glaskörper (des Auges)	VITREOUS BODY (OF THE EYE)	Corps vitré
Gleittisch	SLIDING STAGE	Platine à glissement
Glühbirne	FILAMENT LAMP	Lampe à incandescence
Glühen	INCANDESCENCE	Incandescence
Glühfadenlampe	FILAMENT LAMP	Lampe à incandescence
Glühlampe	FILAMENT LAMP	Lampe à incandescence
Glühwendel	FILAMENT	Filament
Goniometer (Winkelmesser)	GONIOMETER	Goniomètre
Goniometerokular	GONIOMETER EYEPIECE	Oculaire goniométrique
Graufilter	NEUTRAL DENSITY FILTER or NEUTRAL FILTER	Filtre gris; Densité neutre
Greenough-Mikroskop	GREENOUGH MICROSKOPE	Microscope de Greenough
Grobtrieb	COARSE ADJUSTMENT	Mouvement rapide
Größe, radiometrische	RADIOMETRIC QUANTITY	Grandeur énergétique
Großfeldokular	WIDE-FIELD EYEPIECE	Oculaire à grand champ
Halbwertswellenlänge	HALF-PEAK-HEIGHT WAVELENGTH	Longeur d'onde à mi-hauteur
Halo	HALO	Halo
Halogenlampe	HALOGEN LAMP	Lampe à halogène
hängender Tropfen	HANGING-DROP	Goutte pendante
Hauptebene	PRINCIPAL PLANE	Plan principal
Hauptpunkt	PRINCIPAL POINT	Point principal
Hauptschwingungsrichtungen	PRINCIPAL VIBRATION DIRECTIONS	Directions principales de vibration
Hauptstrahl	PRINCIPAL RAY	Rayon principal
Helladaptation	LIGHT ADAPTATION	Adaptation à la lumière
Helladaptationszustand	STATE OF LIGHT ADAPTATION	État d'adapation (à la lumière)
Hellbezugswert	LUMINANCE FACTOR	Facteur de luminance
Hellempfindlichkeitsgrad, spektraler	SPECTRAL LUMINOUS EFFICIENCY (OF THE EYE)	Éfficacité lumineuse relative spectrale
Hellempfindlichkeitskurve, spektrale, des Auges (Vλ-Kurve)	SPECTRAL LUMINOSITY CURVE	Courbe d'éfficacité lumineuse relative spectrale de l'oeil, V (lamda)
Hellfeld	BRIGHT FIELD	Fond clair
Hellfeldbeleuchtung	BRIGHT-FIELD ILLUMINATION	Éclairage en fond clair
Hellfeldbild	BRIGHT-FIELD IMAGE	Image en fond clair
Hellfeldmikroskopie	BRIGHT-FIELD MICROSCOPY	Microscopie en fond clair
Helligkeit	LUMINOSITY; BRIGHTNESS	Luminosité

Heizkammer	WARM CHAMBER	Chambre chaud
Heiztisch	HEATING STAGE	Platine chauffante
Hilfsmikroskop	AUXILIARY TELESCOPE	Viseur de Bertrand
Hohlschliffobjektträger	HANGING DROP SLIDE	Lame creuse
Hohlraumstrahler	BLACK BODY RADIATOR	Corps noir; Radiateur intégral
Homal®	HOMAL® (HOMAL EYEPIECE)	Homal®
homogene Immersion	HOMOGENEOUS IMMERSION	Immersion homogène
homogene Immersion des Objektivs (oder Kondensors)	HOMOGENEOUS IMMERSION OF OBJECTIVE (OR CONDENSER)	Immersion homogène de l'objectif (ou du condenseur)
Hornhaut des Auges	CORNEA	Cornée
Huygensokular	HUYGENIAN EYEPIECE	Oculaire de Huygens
Hypermetropie (Weitsichtigkeit)	HYPERMETROPIA; (FARSIGHTEDNESS)	Hypermétropie
Immersion	IMMERSION	Immersion
Immersion, homogene	HOMOGENEOUS IMMERSION	Immersion homogène
Immersionsansatz	DIPPING CONE	Cône d'immersion
Immersionsflüssigkeit	IMMERSION LIQUID	Liquide d'immersion
Immersionsöl	IMMERSION OIL	Huile d'immersion
Immersionsrefraktometrie	IMMERSION REFRACTOMETRY	Réfractométrie par immersion
Immersionssystem	IMMERSION LENS	Optique à immersion
Indikatrix	INDICATRIX	Ellipsoïde des indices
Infrarot-Mikroskopie	INFRA-RED MICROSCOPY	Microscopie en l'infra-rouge
Infrarotstrahlung	INFRA-RED RADIATION	Rayonnement infra-rouge
inkohärent	INCOHERENT	Incohérent
Integrationsokular	INTEGRATING EYEPIECE	Oculaire de comptage
Integrationstisch	INTEGRATING STAGE	Platine de comptage
Intensität	INTENSITY	Intensité
Intensitätskontrast	PHOTOMETRIC CONTRAST	Contraste photométrique
Interferenz	INTERFERENCE	Interférence
Interferenz, Zweistrahl-	DOUBLE-BEAM INTERFERENCE	Interférence à deux ondes
Interferenzfarbe	INTERFERENCE COLOUR	Couleurs interférentielles
Interferenzfilter	INTERFERENCE FILTER	Filtre interférentiel
Interferenzkontrast	INTERFERENCE CONTRAST	Contrast interférentiel
Interferenzmikroskop	INTERFERENCE MICROSCOPE	Micro-interféromètre
Interferenzmikroskopie	INTERFERENCE MICROSCOPY	Microscopie interférentielle
Interferometer, Doppelfokus-	DOUBLE-FOCUS INTERFEROMETER	Interféromètre à deux foyers
Interferometer, Shearing-	SHEARING INTERFEROMETER	Interféromètre à dédoublement latéral
Interferometer, differentielles	DIFFERENTIAL INTERFEROMETER	Interféromètre différentiel
Interferometrie	INTERFEROMETRY	Interférométrie
Interferometrie, Mikroskop-	MICROSCOPE INTERFEROMETRY	Micro-interférométrie
Interstitium	HIATUS	Interstice
Inversion	INVERSION	Inversion de pseudo-relief
Iris (des Auges)	IRIS	Iris
Irisblende	IRIS DIAPHRAGM	Diaphragme iris
ISO	ISO	ISO
Isochromaten	ISOCHROMATIC CURVES	Lignes isochromatiques
Isogyren	ISOGYRES	Isogyres
Isotropie	ISOTROPY	Isotropie
Kammerwasser	AQUEOUS HUMOR	Humeur aqueuse
Kanadabalsam	CANADA BALSAM	Baume du Canada
Kardinalebenen	CARDINAL PLANES	
Kardinalelemente	CARDINAL ELEMENTS	Éléments cardinaux
Kardinalpunkte	CARDINAL POINTS	
Kardinalstrecken	CARDINAL DISTANCES	
Kardioidkondensor	CARDIOID CONDENSER	Condenseur cardioïde
katadioptrisch	CATADIOPTRIC	Catadioptrique
katoptrisch	CATOPTRIC	Catoptrique

Kellnersches Okular	KELLNER EYEPIECE	Oculaire de Kellner
kissenförmige Verzeichnung	PINCUSHION DISTORTION	Distortion en coussinet
Kitt (Feinkitt)	CEMENT	Colle
Knotenpunkte	NODAL POINTS	Points nodeaux
Knotenpunktsebene	NODAL PLANE	Plan nodal
Köhlersche Beleuchtung	KÖHLER ILLUMINATION	Éclairage de Köhler
kohärent	COHERENT	Cohérent
Kohärenz	COHERENCE	Cohérence
Kohärenz, partielle	PARTIAL COHERENCE	Cohérence partielle
Kohärenzbedingung	COHERENCE CONDITION	Condition de cohérence
Kohärenzgrad	DEGREE OF COHERENCE	Degré de cohérence
Kollektor	COLLECTOR	Lentille collectrice
Kollektorblende	COLLECTOR DIAPHRAGM	Diaphragme de lampe
Kollimation	COLLIMATION	Collimation
Kollimator	COLLIMATOR	Collimateur
kollimieren	COLLIMATE	Collimater
Koma	COMA	Coma; Aigrette
Kompensator	COMPENSATOR	Compensateur
kompensative Wellenlänge	COMPLEMENTARY DOMINANT WAVELENGTH	Longeur d'onde dominante complémentaire
Kompensationsokular	COMPENSATING EYEPIECE	Oculaire compensateur
Kompressarium	COMPRESSARIUM	Chambre de compression
Kondensor	CONDENSER	Condenseur
Kondensor mit ausklapp-barer Frontlinse	SWING-OUT TOP LENS CONDENSER	Condenseur à lentille frontale escamotable
Kondensor, achromatisch-aplanatischer	ACHROMATIC-APLANATIC CONDENSER	Condenseur aplanétique et achromatique
Kondensor, pankratischer	PANCRATIC CONDENSER	Condenseur pancratique
Kondensorblende	CONDENSER DIAPHRAGM	Diaphragme du condenseur
konfokale Abbildung	CONFOCAL IMAGING MODE	Imagerie confocale
konjugiert	CONJUGATE	Conjugué
konjugierte Strahlenanteile	CONJUGATE PARTS OF A RAY	
konjugierte Ebenen	CONJUGATE PLANES	Plans conjugués
konjugierte Punkte	CONJUGATE POINTS	Points conjugués
konoskopische Beobachtung; (Konoskopie)	CONOSCOPIC OBSERVATION (or CONOSCOPY)	Observation conoscopique (ou Conoscopie)
konoskopisches Bild; Achsenbild	CONOSCOPIC (INTERFERENCE FIGURE)	Figure d'axes
kontinuierliches Spektrum	CONTINUOUS SPECTRUM	Spectre continu
Kontrast	CONTRAST	Contraste
Kontrast, photometrischer	PHOTOMETRIC CONTRAST	Contraste photométrique
Kontrast, physiologischer	PHYSIOLOGICAL CONTRAST	Contraste physiologique
Kontrastfilter	CONTRAST FILTER	Filtre de contraste
Konvergenz	CONVERGENCE	Convergence
Konvergenzwinkel	CONVERGENCE ANGLE	Angle de convergence
Konvergenzpunkt	CONVERGENCE POINT	Point de convergence
konvergentes Strahlenbündel	CONVERGING RAY BUNDLE	Faisceau de rayons convergents
Konversionsfilter (Farb-)	COLOUR-CONVERSION FILTER	Filtre de conversion de température de couleur
Korrektion	CORRECTION	Correction
Korrektionsfassung	CORRECTION COLLAR	Bague de correction
Korrektionsklasse	CORRECTION CLASS	Classe de correction
Korrektion auf die Bildweite	CORRECTION FOR IMAGE DISTANCE	Distance image de correction
Kreuztisch	MECHANICAL STAGE	Platine à mouvements croisés; Surplatine
kristalloptische Achse	CRYSTAL OPTICAL AXIS	Axe optique d'un cristal
kritischer Winkel	CRITICAL ANGLE	Angle limite
kritische Beleuchtung	ILLUMINATION, SOURCE-FOCUSED	Éclairage critique
Kronglas	CROWN GLASS	Crown
Kühlkammer	COLD CHAMBER	Chambre froid
Kühltisch	COOLING STAGE	Platine réfrigérante
Kulturkammer	CULTURE CHAMBER	Chambre de culture
Kurzpaßfilter	SHORT-WAVE-PASS FILTER	Filtre passe-bas
Kurzsichtigkeit	NEARSIGHTEDNESS	Myopie

Lambda-Viertel-Plättchen; ¼λ-Plättchen	QUARTER WAVE PLATE	Lame quart d'onde
Lambert–Beersches Gesetz	BEER–LAMBERT LAW	Loi de Beer–Lambert
Lampe	LAMP; BULB	Lampe
Lampenkolben	BULB; ENVELOPE	Ampoule
Langpassfilter	LONG-WAVE-PASS FILTER	Filtre passe haut
Laser	LASER	Laser
Lateralvergrößerung	LATERAL MAGNIFICATION	Grandissement transversal
Leervergrößerung	EMPTY MAGNIFICATION	Grossissement vide
Lemniskate	LEMNISCATE	Lemniscate
Leuchtdichte	LUMINANCE	Luminance
Leuchtfeld	ILLUMINATED FIELD	Champ éclairé
Leuchtfeldblende	ILLUMINATED FIELD DIAPHRAGM	Diaphragme de champ
Licht (sichtbare Strahlung)	LIGHT (VISIBLE RADIATION)	Lumière (Rayonnement visible)
Licht, abgebeugtes	DIFFRACTED LIGHT	Lumière diffractée
Licht, direktes	DIRECT LIGHT	Lumière directe
Licht, elliptisch polarisiertes	ELLIPTICALLY POLARIZED LIGHT	Lumière polarisée elliptique
Licht, linear polarisiertes	PLANE-POLARIZED LIGHT	Lumière polarisée rectiligne
Licht, natürliches	NATURAL LIGHT	Lumière naturelle
Licht, polarisiertes	POLARIZED LIGHT	Lumière polarisée
Licht, weißes	WHITE LIGHT	Lumière blanche
Licht, zirkular polarisiertes	LIGHT, CIRCULARLY-POLARIZED	Lumière polarisée circulaire
Lichtbündel	LIGHT BEAM	Faisceaux lumineux
Lichtbündel,einfallendes (oder auffallendes)	INCIDENT BEAM	Faisceaux incident
Lichtfilter	LIGHT FILTER	Filtre
Lichtmikroskop	LIGHT MICROSCOPE	Microscope optique
Lichtmikroskopie	LIGHT MICROSCOPY	Microscopie (optique)
Lichtquelle	SOURCE	Source
Lichtquelle, ausgedehnte	EXTENDED SOURCE	Source étendue
Lichtquelle, primäre	PRIMARY SOURCE	Source primaire
Lichtquelle, punktförmige	POINT SOURCE	Source ponctuelle; Point lumineux
Lichtquelle, stellvertretende	SECONDARY SOURCE (SUBSTITUTE)	Source secondaire
Lichtstärke	LUMINOUS INTENSITY	Intensité lumineuse
Lichtstrom	LUMINOUS FLUX	Flux lumineux
Lichtvektor	LIGHT VECTOR	Vecteur électrique; Vecteur de Fresnel; Vibration lumineuse
Lichtweg	LIGHT PATH	Chemin optique
Lieberkühnspiegel	LIEBERKÜHN (LIEBERKÜHN REFLECTOR)	Miroir de Lieberkühn
linear polarisiertes Licht	PLANE-POLARIZED LIGHT	Lumière polarisée rectiligne
Linearvergrößerung	LINEAR MAGNIFICATION	Grandissement latéral
Linienspektrum	LINE SPECTRUM	Spectre de raies
Linse	LENS	Lentille; Objectif
Linse, asphärische	ASPHERICAL LENS	Lentille asphérique
Linse, Amici-Bertrand- (Bertrand-)	BERTRAND (AMICI-BERTRAND) LENS	Lentille de Bertrand (Amici-Bertrand)
Linse, dicke	THICK LENS	Lentille épaisse
Linse, dünne	THIN LENS	Lentille mince
Linse, oberste (Frontlinse) des Kondensors	TOP LENS	Lentille frontal du condenseur
Linse des Auges	LENS OF THE EYE; CRYSTALLINE LENS	Cristallin
Linsenfassung	LENS MOUNT	Monture de lentille
Linsenpapier	LENS TISSUE (LENS PAPER)	Papier à nettoyer les lentilles
Linsenscheitel	VERTEX	Pôle
Luftbild	AERIAL IMAGE	Image aérienne
Lumen	LUMEN	Lumen
Lumineszenz	LUMINESCENCE	Luminescence

Lupe	MAGNIFIER; LOUPE	Loupe
Lux	LUX	Lux
Makrophotographie	PHOTOMACROGRAPHY	Macrophotographie
Maßstabs-Strich	SCALE BAR	Échelle
Maßstabszahl des Objektivs	MAGNIFICATION POWER (NUMBER) OF OBJECTIVE	Grandissement de l'objectif
Mattscheibe	GROUND GLASS	Verre dépoli
Maximalwert des spektralen photometrischen Strahlungs-äquivalents	MAXIMUM SPECTRAL LUMINOUS EFFICACY (OF THE EYE)	Efficacité lumineuse spectrale maximale
mechanische Tubuslänge	MECHANICAL TUBELENGTH	Longeur mécanique de tube
Medium	MEDIUM	Milieu
Meniskus; Meniskuslinse	MENISCUS LENS	Lentille ménisque
mesopisches Sehen	MESOPIC VISION	Vision crépusculaire (mésopique)
Meßmikroskop	MEASURING MICROSCOPE	Microscope de mesures
Meßokular	MICROMETER EYEPIECE	Oculaire micrométrique
Mikrodensitometer	MICRODENSITOMETER	Microdensitomètre
Mikrohärteprüfer	MICROHARDNESS TESTER	Microduromètre
Mikrointerferometer	MICROINTERFEROMETER	Microinterféromètre
Mikrokopie	MICROPHOTOGRAPHY	Microphotographie
Mikromanipulator	MICROMANIPULATOR	Micromanipulateur
Mikrometer	MICROMETER	Micromètre
Mikrometer, μm	MICROMETRE, μm	Micromètre, μm
Mikrometerokular	MICROMETER EYEPIECE	Oculaire micrométrique
Mikrophotographie	PHOTOMICROGRAPHY	Photomicrographie
Mikrophotographische Aufnahme	MICROGRAPH	Photomicrographie
mikrophotographische Feldblende	PHOTOMICROGRAPHIC FIELD DIAPHRAGM	Diaphragme du champ photographié
Mikrophotographische Vergrößerung	PHOTOMICROGRAPHIC MAGNIFICATION	Grandissement photographique; Échelle
Mikrophotometer	MICROPHOTOMETER	Microphotomètre
Mikrophotometrie	MICROPHOTOMETRY	Microphotométrie
Mikroprojektor	MICROPROJECTOR	Microscope de projection
Mikroskop	MICROSCOPE	Microscope
Mikroskop, Binokular-	BINOCULAR MICROSCOPE	Microscope binoculaire; Loupe binoculaire
Mikroskop, Fluoreszenz-	FLUORESCENCE MICROSCOPE	Microscope pour fluorescence
Mikroskop, Interferenz-	INTERFERENCE MICROSCOPE	Microscope interférentiel
Mikroskop, einfaches (einstufiges)	SIMPLE MICROSCOPE	Microscope simple
Mikroskop, monokulares	MONOCULAR MICROSCOPE	Microscope avec tube monoculaire
Mikroskop, umgekehrtes	INVERTED MICROSCOPE	Microscope inversé
Mikroskop, zusammengesetztes	COMPOUND MICROSCOPE	Microscope (composé)
Mikroskopfluorometrie	MICROSCOPE FLUORIMETRY	Microfluorométrie
Mikroskopfuß	MICROSCOPE BASE	Pied du microscope
Mikroskopie	MICROSCOPY	Microscopie
Mikroskopie, quantitative	QUANTITATIVE MICROSCOPY	Microscopie quantitative
Mikroskopie, Infrarot-	INFRA-RED MICROSCOPY	Microscopie en infrarouge
Mikroskopie, Phasenkontrast	PHASE-CONTRAST MICROSCOPY	Microscopie en contraste de phase
Mikroskopie, Reliefkontrast	RELIEF-CONTRAST MICROSCOPY	Microscopie en contrast de relief
Mikroskopierleuchte	MICROSCOPE LAMP	Lampe de microscope
Mikroskopinterferometrie	MICROSCOPE INTERFEROMETRY	Micro-Interférométrie
mikroskopisch (mikroskopisch klein)	MICROSCOPIC	Microscopique
mikroskopisch (im Zusammenhang mit dem Mikroskop)	MICROSCOPICAL	Microscopique
Mikroskopphotometer	MICROSCOPE PHOTOMETER	Microphotomètre
Mikroskopphotometrie	MICROSCOPE PHOTOMETRY	Microphotometrie
Mikroskopspiegel	MICROSCOPE MIRROR	Miroir d'éclairage du microscope

APPENDIX IV

Mikroskopstativ	STAND (MICROSCOPE STAND)	Statif du microscope
Mikroskoptubus	BODY TUBE	Tube du microscope
Mikrospektralphotometer	MICROSPECTROPHOTOMETER	Microspectrophotomètre
Mikrospektralphotometrie	MICROSPECTROPHOTOMETRY	Microspectrophotométrie
Mikrotom	MICROTOME	Microtome
Mired	MIRED	Mired
Mischstrahlung	COMPLEX RADIATION; (COMPOUND RADIATION)	Rayonnement polychromatique
Modalanalyse	MODALANALYSIS	
Modulationskontrast	MODULATION CONTRAST	Contraste de modulation
Molekularbewegung, Brownsche	BROWNIAN MOVEMENT	Mouvement brownien
Monochromasie (Farbenblindheit)	MONOCHROMATISM	Achromatopsie
Monochromat	MONOCHROMAT	Monochromat
monochromatische Strahlung	MONOCHROMATIC RADIATION	Rayonnement monochromatique
Monochromator	MONOCHROMATOR	Monochromateur
monokular	MONOCULAR	Monoculaire
monokulares Mikroskop	MONOCULAR MICROSCOPE	Microscope monoculaire
Monokulartubus	MONOCULAR TUBE	Tube monoculaire
Mouches volantes	*MUSCAE VOLITANTES*	Mouches volantes
Myopie (Kurzsichtigkeit)	MYOPIA (NEARSIGHTEDNESS)	Myopie
Nachbild	AFTER-IMAGE	Impression rémanente
Nachtsehen	NIGHT VISION	Vision de nuit; Vision scotopique
Nahpunkt (des Auges)	NEAR POINT (OF THE EYE)	Punctum proximum
Nahpunktsabstand	NEAR POINT DISTANCE (NEAREST DISTANCE OF DISTINCT VISION)	Distance du punctum proximum; Distance minimale de vision distincte
Negativlinse	NEGATIVE LENS	Lentille divergente
Netzhaut (des Auges); Retina	RETINA	Rétine
Netzhaut, Stäbchen der	RODS (OF THE RETINA)	Batonnets de la rétine
Netzhaut, Zapfen der	CONES (OF THE RETINA)	Cônes de la rétine
Netzhautbild	RETINAL IMAGE	Image rétienne
Netzhautgrube (Fovea centralis)	FOVEA	Fovea
Neutralfilter	NEUTRAL DENSITY FILTER	Filtre gris; Densité neutre
Newtonsche Interferenzfarben	NEWTON'S SCALE OF COLOURS	Échelle des teintes de Newton
Nicolsches Prisma	POLARIZING PRISM (NICOL PRISM)	Nicol
nicht querdisparates Tiefensehen	NON-DISPARATE SPATIAL VISION	Vision monoculaire du relief
Nomarski, differentieller Interferenzkontrast nach	NOMARSKI DIFFERENTIAL INTERFERENCE CONTRAST	Contrast interférentiel différentiel de Nomarski
Normalstellung	NORMAL POSITION	Position avec axes parallèles
Normfarbtafel	CHROMATICITY DIAGRAM	Triangle des couleurs; Diagramme de chromaticité
Normfarbwertanteile	CHROMATICITY CO-ORDINATES	Coordonnées tri-chromatiques
Normfarbwerte (trichromatische Farbmaßzahlen)	TRISTIMULUS COLOUR VALUES	Composantes tri-chromatiques
Normfarbwertkoordinaten	CHROMATICITY CO-ORDINATES	Coordonnées tri-chromatiques
numerische Apertur	NUMERICAL APERTURE	Ouverture numérique
Nutzeffekt, visueller	LUMINOUS EFFICIENCY	Efficacité lumineuse relative
oberste Linse (Frontlinse) des Kondensors	TOP LENS (OF CONDENSER)	Lentille frontale du condenseur
Objekt	OBJECT	Objet
Objekt, poliertes	POLISHED SECTION	Échantillon poli; Section polie
Objekt, selbstleuchtendes	SELF-LUMINOUS OBJECT	Objet lumineux par lui-même

Objekt-Zwischenbild Abstand	OBJECT TO PRIMARY (INTERMEDIATE) IMAGE DISTANCE	Distance entre l'objet et l'image
Objektebene	OBJECT PLANE	Plan objet
Objektfeld	OBJECT FIELD	Champ objet
Objektfinder	OBJECT FINDER	Lame de repérage
Objektführer	MECHANICAL STAGE	Surplatine
Objektmarkierer	OBJECT MARKER	Marqueur d'objets
Objektmikrometer	STAGE MICROMETER	Micromètre-objet
Objektiv	OBJECTIVE	Objectif
Objektiv, achromatisches	ACHROMATIC OBJECTIVE	Objectif achromatique
Objektiv, apochromatisches	APOCHROMATIC OBJECTIVE	Objectif apochromatique
Objektiv, auf unendlich korrigiertes	INFINITY-CORRECTED OBJECTIVE	Objectif corrigé à l'infini
Objektiv, *Fluorit-	SEMI-APOCHROMAT	Objectif semi-apochromatique
Objektiv(s), freier Objektabstand eines	FREE WORKING DISTANCE (OF THE OBJECTIVE)	Distance frontale d'un objectif
Objektiv mit Federfassung	SPRING-LOADED OBJECTIVE	Objectif en monture rétractable
Objektiv, mit großem freiem Arbeitsabstand	LONG WORKING DISTANCE OBJECTIVE	Objectif à grande distance frontale
Objektiv, semiapochromatisches	SEMI-APOCHROMATIC OBJECTIVE	Objectif semi-apochromatique
Objektiv, spannungsfreies	STRAIN-FREE OBJECTIVE	Objectif sans tension
Objektiv-Anlagefläche	LOCATING-FLANGE OF OBJECTIVE	Épaulement d'un objectif
Objektiv-Anlagefläche (am Objektivrevolver)	OBJECTIVE-LOCATING SURFACE (OF THE NOSEPIECE)	Face d'appui de l'objectif (sur le revolver)
Objektiv-Anschraubgewinde	SCREW THREAD, FOR OBJECTIVE	Filetage d'un objectif
Objektiv-Standardgewinde	RMS THREAD	Filetage RMS
Objektiv-Zwischenbild-Abstand	OBJECTIVE TO PRIMARY (INTER MEDIATE) IMAGE DISTANCE	Distance entre la face d'appui de l'objectif et l'image intermédiaire
Objektivhalter	NOSEPIECE	Porte-objectif
Objektivhalter, zentrierbarer	CENTRING NOSEPIECE	Porte-objectif centrable
Objektivrevolver	REVOLVING NOSEPIECE	Revolver
Objektivwechsler	NOSEPIECE	Porte-objectif
Objektivwechsler, zentrierbarer	CENTRING NOSEPIECE	Porte objectif centrable
Objektraum	OBJECT SPACE	Espace objet
objektseitige numerische Apertur (eines Mikroskopobjektivs)	OBJECT-SIDE NUMERICAL APERTURE (OF A MICROSCOPE OBJECTIVE)	Ouverture numérique (d'un objectif de microscope)
Objekttisch (des Mikroskops)	STAGE; MICROSCOPE STAGE	Platine (de microscope)
Objekttisch, drehbarer	ROTATING STAGE	Platine tournante
Objekttisch, zentrierbarer	CENTRING STAGE	Platine centrable
Objektträger	SLIDE; MICROSCOPE SLIDE	Lame porte-objet
Objektweite	OBJECT DISTANCE	Distance objet
Öffnungsfehler	APERTURE ABERRATION	Aberration sphérique
Öffnungswinkel	ANGULAR APERTURE	Ouverture angulaire
Okular einstellbares	FOCUSABLE EYEPIECE	Oculaire à mise au point réglable
Okular mit Vorderblende	EXTERNAL-DIAPHRAGM EYEPIECE	Oculaire positif
Okular mit Zwischenblende	INTERNAL-DIAPHRAGM EYEPIECE	Oculaire negatif
Okular, orthoskopisches	ORTHOSCOPIC EYEPIECE	Oculaire orthoscopique; (de Kellner)
Okularschraubenmikrometer	MICROMETER SCREW EYEPIECE; FILAR EYEPIECE	Oculaire à vis micrométrique
Okularvergrößerung	MAGNIFYING POWER OF EYEPIECE	Grossissement commercial (ou conventionnel) de l'oculaire
Okularstrichplatte	EYEPIECE-GRATICULE	Graticule oculaire
Okularanlagefläche des Beobachtungstubus	EYEPIECE-LOCATING SURFACE (OF VIEWING TUBE)	Face d'appui de l'oculaire (sur le tube)
Opalglas	OPAL GLASS	Verre opale
Optik, geometrische	GEOMETRICAL OPTICS	Optique géométrique
optisch einachsig	OPTICALLY UNIAXIAL	Uniaxe

optisch zweiachsig	OPTICALLY BIAXIAL	Biaxe
optische Achse	OPTICAL AXIS	Axe optique
optische Aktivität	OPTICAL ACTIVITY	Activité optique; Pouvoir rotatoire
optische Anisotropie	OPTICAL ANISOTROPY	Anisotropie optique
optische Dicke (Weglänge)	OPTICAL PATHLENGTH; (OPTICAL DISTANCE)	Chemin optique
optische Drehung	OPTICAL ROTATION	Polarisation rotatoire
optische Tubuslänge	OPTICAL TUBELENGTH	Longueur optique de tube
optische Weglänge	OPTICAL PATHLENGTH; (OPTICAL DISTANCE)	Chemin optique
optische Weglängendifferenz	OPTICAL PATHLENGTH DIFFERENCE	Différence de chemin optique
optischer Achsenwinkel	OPTIC AXIAL ANGLE	Angle des axes optiques (d'un cristal biaxe)
ordentlicher Strahl	ORDINARY RAY	Rayon ordinaire
orthoskopische Beobachtung	ORTHOSCOPIC OBSERVATION	Observation orthoscopique
orthoskopisches Okular	ORTHOSCOPIC EYEPIECE	Oculaire orthoscopique (de Kellner)
paraxial; achsennah	PARAXIAL	Paraxial
Paraxialstrahl	PARAXIAL RAY	Rayon paraxial
pankratisch	PANCRATIC	Pancratique
Paraboloidkondensor	PARABOLOID CONDENSER	Condenseur parabolique
Parallaxe	PARALLAX	Parallaxe
parallele Polare	PARALLEL POLARS	Polariseurs parallèles
paralleler Strahlengang	PARALLEL RAY PATH	Marche parallèle de rayons
paralleles Strahlenbündel	PARALLEL RAY BUNDLE	Faisceau de rayons parallèles
parfokal (untereinander abgeglichen)	PARFOCAL	Équilibré
partielle Kohärenz	PARTIAL COHERENCE	Cohérence partielle
Phase	PHASE	Phase
Phasengeschwindigkeit	PHASE VELOCITY	Vitesse de phase
Phasenkontrastkondensor	PHASE-CONTRAST CONDENSER	Condenseur pour contraste de phase
Phasenkontrastmikroskopie	PHASE-CONTRAST MICROSCOPY	Microscopie en contraste de phase
Phasenplatte	PHASE-PLATE; RETARDATION PLATE	Lame de phase
Phasenschieber	RETARDATION PLATE	Lame auxiliaire; Lame retard
Phasenverschiebung	PHASE SHIFT	Déphasage
Phasenunterschied	PHASE DIFFERENCE	Différence de phase
Phasenwinkel	PHASE ANGLE	Phase
Phosphoreszenz	PHOSPHORESCENCE	Phosphorescence
Photolumineszenz	PHOTOLUMINESCENCE	Photoluminescence
Photokopf	CAMERA HEAD	Tête photographique
Photometer	PHOTOMETER	Photomètre
Photometerblende (photometrische Blende)	DIAPHRAGM PHOTOMETRIC	Diaphragne du champ photométré
Photometerkopf	PHOTOMETER HEAD	Tête photométrique
Photometertubus	PHOTOMETER HEAD	Tête photométrique
Photometrie	PHOTOMETRY	Photométrie
Photometrie, Mikroskop-	MICROSCOPE PHOTOMETRY	Microphotométrie
photometrisch	PHOTOMETRIC	Photométrique
photometrische Blende	PHOTOMETRIC DIAPHRAGM	Diaphragme du champ photométré
photometrisches Feld	PHOTOMETRIC FIELD	Champ photométré
photometrische Größe	PHOTOMETRIC QUANTITY	Grandeur photométrique
Photon	PHOTON	Photon
photopisches Sehen (Tagessehen)	PHOTOPIC VISION (DAYLIGHT VISION)	Vision photopique
photometrischer Kontrast	PHOTOMETRIC CONTRAST	Contraste
Phototubus	CAMERA TUBE	Tube photographique
physiologischer Kontrast	PHYSIOLOGICAL CONTRAST	Contraste physiologique
Planckscher Strahler	BLACK BODY RADIATOR	Corps noir; Radiateur intégral

Planobjektiv	FLAT-FIELD OBJECTIVE	Objectif à champ plan
Planokular	FLAT-FIELD EYEPIECE	Oculaire à champ plan
Plättchen, Rot 1. Ordnung	FIRST-ORDER RED PLATE	Lame à teinte sensible du 1° ordre
Plättchen, ½λ (½ Lambda)	HALF-WAVE PLATE; ½λ PLATE	Lame demi-onde
Plättchen, ¼λ (¼ Lambda)	QUARTER-WAVE PLATE; ¼λ PLATE	Lame quart d'onde
Pleochroismus	PLEOCHROISM	Pléochroïsme
Point Counter	POINT COUNTER	"Pointcounter"
Polar (Polarisator)	POLAR	Polariseur
Polare (Polarisatoren), gekreuzte	CROSSED POLARS	Polariseurs croisés
Polare (Polarisatoren), parallele	PARALLEL POLARS	Polariseurs parallèles
Polare (Polarisatoren), ungekreuzte	UNCROSSED POLARS	Polariseurs décroisés
Polarisationsbild	POLARIZATION FIGURE	Figure de convergence
Polarisationsebene	POLARIZATION PLANE	Plan de polarisation
Polarisationsfigur (Achsenbild)	POLARIZATION FIGURE	Figure de convergence
Polarisationsfilter	POLARIZING FILTER	Filtre polarisant
Polarisationsgrad	DEGREE OF POLARIZATION	Taux de polarisation
Polarisationsinterferenz	POLARIZING INTERFERENCE	Interférences, par double réfraction, en lumière polarisée
Polarisationsmikroskop	POLARIZED LIGHT MICROSCOPE	Microscope polarisant
Polarisationsmikroskopie	POLARIZED LIGHT MICROSCOPY	Microscopie polarisante
Polarisationsprisma	POLARIZING PRISM	Polariseur; Prism de . . . . (Nicol, Foucault,Glacebrook, Glan etc)
Polarisationsrichtung (der elektromagnetischen Strahlung)	POLARIZATION DIRECTION (OF ELECTROMAGNETIC RADIATION)	Orientation du plan de polarisation (d'un onde éléctromagnétique)
Polarisationszustand	POLARIZATION STATE	État de polarisation
Polarisator	POLARIZER	Polariseur
Polaroid®	POLAROID®	Polaroid®
Polaroidfilm®	POLAROID® FILM	Film Polaroid®
polierter Anschliff	POLISHED SECTION	Échantillon poli; Section polie
Positivlinse (Sammellinse)	POSITIVE LENS	Lentille convergente
Präparat	SPECIMEN; PREPARATION	Échantillon; Objet
Präparation	PREPARATION	Préparation
Präpariermikroskop	DISSECTING MICROSCOPE	Microscope de dissection
Presbyopie (Alters-sichtigkeit	PRESBYOPIA	Presbytie
primäres Beugungsbild (Beugungsfigur; Inter-ferenzbild)	PRIMARY DIFFRACTION PATTERN (OR IMAGE)	Spectre de diffraction de l'objet; Spectre des fréquences spatiales
primäre Lichtquelle	PRIMARY SOURCE	Source primaire de lumière
Prisma	PRISM	Prisme
Prisma, Nicolsches	NICOL PRISM	Prisme de Nicol
Probe, zu untersuchende	SAMPLE	Échantillon
Projektionskopf	PROJECTION HEAD	Tête de projection
Projektionssystem	PROJECTION LENS	Systéme optique de projection
Projektionsokular	PROJECTION EYEPIECE	Oculaire de projection
Projektiv	PROJECTION LENS	Systéme optique de projection
prozentualer Anteil (einer Komponente am Gesamtbe-stand)	MODE	
Punkte, konjugierte	CONJUGATE POINTS	Points conjugués
Pseudoskopie	PSEUDOSCOPY	Pseudoscopie
Punktförmige Lichtquelle	POINT SOURCE	Source ponctuelle; Point lumineux
Pupille	PUPIL	Pupille

Pupillenabstand	INTERPUPILLARY DISTANCE	Écart d'yeux; Écart interpupillaire
Purkinje-Phänomen	PURKINJE SHIFT	Phénomène de Purkinje
Purpurgerade	PURPLE LINE	Ligne des pourpres
quantitative Mikroskopie	QUANTITATIVE MICROSCOPY	Microscopie quantitative
Quarzkeil	QUARTZ WEDGE	Biseau de Quartz: Coin de Quartz
Quecksilberdampflampe	MERCURY ARC LAMP	Lampe à vapeur de mercure
Queraberration, chromatische	LATERAL CHROMATIC ABERRATION	Chromatisme de grandeur (latéral)
Radiometer (Strahlungsmesser)	RADIOMETER	Radiomètre
Radiometrie	RADIOMETRY	Radiométrie
radiometrisch	RADIOMETRIC	Radiométrique
radiometrische Größe	RADIOMETRIC QUANTITY	Grandeur énergétique
Ramsdenscher Kreis	RAMSDEN DISC	Anneau de Ramsden; Circle oculaire
Ramsden-Okular	RAMSDEN EYEPIECE	Oculaire de Ramsden
Randkontrast	MARGINAL CONTRAST	Contraste marginal
Randstrahl	MARGINAL RAY	Rayon marginal
Raster	RASTER	Trame
räumlich	SPATIAL	Spatial
räumliches Sehen	SPATIAL VISION	Vision du relief
Raum- (als Präfix)	SPATIAL	Spatial
Raumwinkel	SOLID ANGLE	Angle solide
reelles Bild	REAL IMAGE	Image réelle
reelles Zwischenbild	INTERMEDIATE (PRIMARY) IMAGE	Image intermédiaire
Refraktometrie	REFRACTOMETRY	Réfractométrie
Regenbogenhaut (Iris)	IRIS	Iris
Reisemikroskop	PORTABLE MICROSCOPE; FIELD MICROSCOPE	Microscope de voyage
relative Dispersion	RELATIVE DISPERSION	Pouvoir dispersif
Relief	RELIEF	Relief
Reliefkontrast	RELIEF CONTRAST	Contraste de relief
Reliefkontrast-Mikroskopie	RELIEF-CONTRAST MICROSCOPY	Microscopie à contraste de relief
Reflektor	REFLECTOR	Prisme (Miroir) (Lame) d'éclairage
Reflexion	REFLECTION	Réflexion
Reflexion, diffuse	DIFFUSE REFLECTION	Réflexion diffuse
Reflexion, spiegelnde (reguläre Reflexion)	SPECULAR REFLECTION (REGULAR REFLECTION)	Réflexion spéculaire
Reflexionsdrehung	REFLECTION ROTATION	Rotation du plan de polarisation à la réflexion
Reflexionsgrad	REFLECTANCE	Facteur de réflexion
Reflexionspleochroismus	REFLECTION PLEOCHROISM	Pléochroïsme de réflexion
Reflexionswinkel	ANGLE OF REFLECTION	Angle du réflexion
Retina (Netzhaut des Auges)	RETINA	Rétine
ringförmige Beleuchtung	ANNULAR ILLUMINATION	Éclairage annulaire
Ringleuchte	RING LIGHT	Source circulaire
Rot 1. Ordnung	FIRST-ORDER RED	Pourpre teinte sensible du 1° ordre
Rot 1. Ordnung-Plättchen	FIRST-ORDER RED PLATE	Lame d'onde (teinte sensible) du 1° ordre
Sammellinse	CONVERGING LENS	Lentille convergente
Sammelwirkung einer Linse	COLLECTING POWER OF A LENS	Convergence d'une lentille
Scanning-Mikroskop	SCANNING MICROSCOPE	Microscope à balayage
Scanning-Mikroskop, optisches	SCANNING OPTICAL MICROSCOPE	Microscope optique à balayage
Scanning-Tisch	SCANNING STAGE	Platine à balayage
Schärfentiefe	DEPTH OF FIELD	Profondeur de champ
Schärfentiefe (im Bildraum)	DEPTH OF FOCUS (DEPTH OF SHARPNESS IN IMAGE SPACE)	Profondeur de foyer

Schärfentiefe (im Objektraum)	DEPTH OF FIELD (DEPTH OF SHARPNESS IN OBJECT SPACE)	Profondeur de champ
Scheibe	SCREEN	Écran
Scheitelpunktswellenlänge	PEAK WAVELENGTH	Longeur d'onde du pic
schiefe Auslöschung	OBLIQUE EXTINCTION	Extinction oblique
schiefe Beleuchtung, (einseitig)	UNILATERAL OBLIQUE ILLUMINATION	Éclairage oblique
Schielen (Strabismus)	STRABISMUS	Strabisme
Schirm (Auffangschirm)	SCREEN	Écran
Schmalbandfilter	NARROW-BAND-PASS FILTER (OR NARROW BAND FILTER)	Filtre à bande étroite
Schnittweite	INTERSECTION DISTANCE	Distance frontale
Schraubenmikrometer	MICROMETER SCREW	Micromètre
schwarzer Körper	BLACK BODY	Corps noir
schwarze Strahlung	BLACK-BODY RADIATION	Rayonnement du corps noir
schwarzer Strahler	BLACK-BODY RADIATOR	Corps noir; Radiateur intégral
Schwingungsebene	VIBRATION PLANE	Plan de vibration
Schwingungsebene (der elektromagnetischen Strahlung)	VIBRATION PLANE (OF ELECTROMAGNETIC RADIATION)	Plan de vibration (du rayonnement électro-magnétique)
Schwingungsrichtung	VIBRATION DIRECTION	Direction de vibration
Sehachse (des Auges)	OPTIC AXIS (VISUAL AXIS) (OF THE EYE)	Axe visuel (de l'oeil)
Sehen, mesopisches (Dämmerungssehen)	MESOPIC VISION (OR TWILIGHT VISION)	Vision crépusculaire (mésopique)
Sehen, photopisches (Tagessehen)	PHOTOPIC VISION (OR DAYLIGHT VISION)	Vision de jour (photopique
Sehen, räumliches	SPATIAL VISION	Vision du relief
Sehen, skotopisches (Nachtsehen)	SCOTOPIC VISION (OR NIGHT VISION)	Vision de nuit (scotopique)
Sehfeld	FIELD OF VIEW	Champ d'observation
Sehfeldzahl	FIELD-OF-VIEW NUMBER	Indice de champ
Sehfeldblende	VISUAL FIELD DIAPHRAGM	Diaphragme de champ de l'oculaire
Sehorgan	VISUAL ORGAN	Système visuel
Sehschärfe	VISUAL ACUITY	Acuité visuelle
Sekundärfluoreszenz	SECONDARY FLUORESCENCE	Fluorescence induite (secondaire)
Selbstleuchter	SELF-LUMINOUS OBJECT	Objet lumineux par lui-même
Semiapochromat	SEMI-APOCHROMAT	Semi-apochromat
Senkrechte	NORMAL	Normale; Perpendiculaire
Shearing-Interferometer	SHEARING INTERFEROMETER	Interféromètre à dédoublement latéral
sichtbare Strahlung	VISIBLE RADIATION	Radiation visible
Sinusbedingung	SINE CONDITION	Condition de sinus
skotopisches Sehen (Nachtsehen)	SCOTOPIC VISION (NIGHT VISION)	Vision de nuit (scotopique)
Snelliussches Brechungs-gesetz	SNELL'S LAW	Loi de Descartes
spektral	SPECTRAL	Spectral
spektrale Dichte (spektrale Konzentration)	SPECTRAL CONCENTRATION (OR SPECTRAL DENSITY)	Densité spectrale
spektrale Hellempfindlich-keitskurve; Vλ-Kurve	SPECTRAL LUMINOSITY CURVE	Courbe d'efficacité lumineuse relative spectrale de l'oeil; V (lamda)
spektraler Hellempfindlich-keitsgrad	SPECTRAL LUMINOUS EFFICIENCY (OF THE EYE)	Efficacité lumineuse relative spectrale de l'oeil
Spektralfarbenzug	SPECTRUM LOCUS	Spectrum locus
Spektralphotometer	SPECTROPHOTOMETER	Spectrophotomètre
Spektralphotometrie	SPECTROPHOTOMETRY	Spectrophotométrie
Spektrum	SPECTRUM	Spectre
Spektrum, kontinuierliches	CONTINUOUS SPECTRUM	Spectre continu
Spektrum, Linien-	LINE SPECTRUM	Spectre de raies
Spektrum, sekundäres	SECONDARY SPECTRUM	Spectre secondaire
Spektrum, tertiäres	TERTIARY SPECTRUM	

Sperrfilter	BARRIER FILTER	Filtre de blocage; Filtre d'arrêt
spezifische Lichtausstrahlung (an einem Punkt einer Fläche)	LUMINOUS EXITANCE (AT A POINT ON A SURFACE)	Exitance lumineuse (en un point d'une surface)
spezifische Ausstrahlung (an einem Punkt einer Fläche)	RADIANT EXITANCE (AT A POINT ON A SURFACE)	Exitance énergétique (en un point d'une surface)
spezifische Drehung	SPECIFIC ROTATION	Rotation spécifique
spezifisches Brechungsinkrement	SPECIFIC REFRACTION INCREMENT	
sphärische Aberration	SPHERICAL ABERRATION	Aberration sphérique
spiegelnde Reflexion	SPECULAR REFLECTION	Réflexion spéculaire
Spiegelobjektiv	REFLECTING OBJECTIVE	Objectif à miroirs
Spiegeloptik-Mikroskop	REFLECTING MICROSCOPE	Microscope à miroirs
Spiegelung	SPECULAR REFLECTION	Réflexion spéculaire
Stäbchen der Netzhaut	RODS (OF THE RETINA)	Batonnets de la rétine
Stativ (Mikroskopstativ)	STAND (MICROSCOPE STAND)	Statif (de microscope)
Stativbügel	MICROSCOPE LIMB	Potence du microscope
stellvertretende (sekundäre) Lichtquelle	SECONDARY (SUBSTITUTE) SOURCE	Source secondaire
Steradiant	STERADIAN	Stéradian
Stereomikroskop	STEREOMICROSCOPE	Microscope stéréoscopique
Stereologie	STEREOLOGY	Stéréologie
Stereopsis (Stereosehen)	STEREOPSIS	Stereopsie
Stereosehen (Stereopsis)	STEREO VISION (STEREOPSIS)	Vision stéréoscopique (stéréopsie)
*stereoskopisches Sehen	STEREO VISION (STEREOPSIS)	Vision stéréoscopique
Sterntest	STAR TEST	Essai au point lumineux
*Stilb	STILB	Stilb
Störlicht	GLARE	Éblouissement; Lumière parasite
Strabismus (Schielen)	STRABISMUS	Strabisme
Strahl	RAY	Rayon
Strahl, axialer	AXIAL RAY	
Strahl, außerordentlicher	EXTRAORDINARY RAY	Rayon extraordinaire
Strahl, einfallender	INCIDENT RAY	Rayon incident
Strahl, ordentlicher	ORDINARY RAY	Rayon ordinaire
Strahl, paraxialer	PARAXIAL RAY	Rayon paraxial
Strahldichte	RADIANCE	Luminance énergétique
Strahlenanteile	CONJUGATE PORTIONS OF RAYS	
Strahlenbündel	RAY BUNDLE	Faisceau de rayons
Strahlenbündel, divergentes	DIVERGING RAY BUNDLE	Faisceau de rayons divergents
Strahlenbündel, konvergentes	CONVERGING RAY BUNDLE	Faisceau de rayons convergents
Strahlenbündel, paralleles	PARALLEL RAY BUNDLE	Faisceau de rayons parallèles
Strahlenbüschel	RAY PENCIL	Pinceau de rayons
Strahlengang	RAY PATH; LIGHT PATH; OPTICAL PATH	Marche de rayons
Strahlengang, paralleler	PARALLEL RAY PATH	Marche parallèle de rayons
Strahlenraum	RAY SPACE	
Strahler, schwarzer	BLACK BODY RADIATOR	Corps noir; Radiateur intégral
Strahlstärke	RADIANT INTENSITY	Intensité énergétique
Strahlung	RADIATION	Rayonnement
Strahlung eines Hohlraumstrahlers	BLACK BADY RADIATION	Rayonnement du corps noir
Strahlung.,zusammengesetzte	COMPLEX RADIATION	Rayonnement polychromatique
Strahlung, sichtbare	VISIBLE RADIATION	Rayonnement visible
Strahlung, Ultraviolett-	ULTRAVIOLET RADIATION	Rayonnement ultra-violet
Strahlungsäquivalent(s), Maximalwert des spektralen photometrischen	MAXIMUM SPECTRAL LUMINOUS EFFICACY (OF THE EYE)	Efficacité lumineuse spectrale maximale
Strahlungsfluß	RADIANT FLUX	Flux énergétique
Strahlungsmenge	RADIANT ENERGY	Énergie rayonnante

strahlungsphysikalisch	RADIOMETRIC	Radiométrique
Strahlenteiler	BEAM-SPLITTER	Séparateur de faisceaux
Streulicht	STRAY LIGHT	Lumière parasie
Streuscheibe	DIFFUSING SCREEN	Écran diffusant
Strichplatte	GRATICULE	Graticule
Strichplattenokular	GRATICULE EYEPIECE	Oculaire à graticule
submikroskopisch	SUB-MICROSCOPIC	Ultramicroscopique
Subtraktionsstellung	SUBTRACTION POSITION	Position avec axes croisés
symmetrische Auslöschung	SYMMETRICAL EXTINCTION	Extinction symétrique
Tagessehen (photopisches)	DAYLIGHT (PHOTOPIC) VISION	Vision photopique
tertiäres Spectrum	TERTIARY SPECTRUM	
Testobjekt	TEST OBJECT	Objet-test
Tiefensehen, nicht querdisparates	NON DISPARATE SPATIAL VISION	Vision monoculaire du relief
tonnenförmige Verzeichnung	BARREL DISTORTION	Distortion en barillet
Totalreflexion	TOTAL INTERNAL REFLECTION	Réflexion totale
Transmission	TRANSMISSION	Transmission
Transmissionsgrad	TRANSMITTANCE	Facteur de transmission
*Trinokulartubus (binokularer Phototubus)	TRINOCULAR TUBE	Tube trinoculaire
Trockenobjektiv	DRY OBJECTIVE	Objectif à sec
Tubus	TUBE	Tube
Tubusfaktor	TUBE-FACTOR	Facteur de tube
Tubuskopf	TUBE HEAD	Tube droit
Tubuskörper	BODY TUBE	Corps
Tubuslinse	TUBE-LENS	Lentille de tube
Tubuslänge, mechanische	MECHANICAL TUBE-LENGTH	Longeur mécanique de tube
Tubuslänge, optische	OPTICAL TUBE-LENGTH	Longeur optique de tube
Tubuslängen-Korrektions-linse	TUBELENGTH CORRECTION LENS	Lentille de correction de longueur de tube
Tubusträger	LIMB (MICROSCOPE LIMB)	Potence
Übertragungslinse (-system)	RELAY LENS	Véhicule
Ultramikroskopie	ULTRAMICROSCOPY	Ultramicroscopie
Ultraviolett-Mikroskop	ULTRAVIOLET MICROSCOPE	Microscope en Ultra-Violet
Ultraviolett-Mikroskopie	ULTRAVIOLET MICROSCOPY	Microscopie en Ultra-Violet
Ultraviolettstrahlung	ULTRAVIOLET RADIATION	Rayonnement Ultra-Violet
Umfeld	SURROUNDING FIELD	Champ périphérique
umgekehrtes Mikroskop	INVERTED MICROSCOPE	Microscope inversé
ungekreuzte Polare (Polarisatoren)	UNCROSSED POLARS	Polariseurs décroisées
unbunte Farbe	ACHROMATIC COLOUR	Couleur achromatique perçue
unendlich-korrigiertes Objektiv	INFINITY-CORRECTED OBJECTIVE	Objectif corrigé à l'infini
Unterkorrektion	UNDERCORRECTION	Sous-correction
Universaldrehtisch	UNIVERSAL STAGE	Platine universelle
Vergenz (der Fixierlinien)	VERGENCE	Vergence
Vergleichsmikroskop	COMPARISON MICROSCOPE	Microscope de comparaison
Vergleichsokular	COMPARISON EYEPIECE	Tube de comparaison
Vergrößerung	MAGNIFICATION	Grandissement
Vergrößerung des Objektivs (Maßstabszahl)	PRIMARY MAGNIFICATION	Grandissement de l'objectif
Vergrößerung, Angular-	ANGULAR MAGNIFICATION	Grossissement commercial
Vergrößerung, Axial-	AXIAL MAGNIFICATION	Grandissement longitudinal; axial
Vergrößerung, Flächen-	AREAL MAGNIFICATION	
Vergrößerung, Lateral-	LATERAL MAGNIFICATION	Grandissement transversal
Vergrößerung, Leer-	EMPTY MAGNIFICATION	Grossissement vide
Vergrößerung, Linear-	LINEAR MAGNIFICATION	Grandissement transversal
Vergrößerung, mikrophotographische	PHOTOMICROGRAPHIC MAGNIFICATION	Grandissement photomicrographique; Échelle
Vergrößerungsbereich, förderlicher	USEFUL RANGE OF MAGNIFICATION	Domaine des grossissements utiles

Vergrößerungsdifferenz, chromatische	LATERAL CHROMATIC ABERRATION	Chromatisme de grandeur; (latéral)
Vergrößerungsmaßstab	LATERAL MAGNIFICATION	Échelle
Vergrößerungsvermögen	MAGNIFYING POWER	1) Grandissement (if the lens gives a real image) 2) Grossissement (if the lens gives a virtual image for visual observation at infinity)
Vergrößerungswechsler	MAGNIFICATION CHANGER	Changeur de grossissement
Vergütung von Linsenoberflächen	COATING OF LENS SURFACES	Traitement de surface des lentilles
Version (beider Augen)	VERSION (OF THE EYES)	Version (des yeux)
Verteilungstemperatur	DISTRIBUTION TEMPERATURE	Témpérature de répartition
Verzeichnung	DISTORTION	Distortion
Verzeichnung, kissenförmige	PINCUSHION DISTORTION	Distortion en coussinet
Verzeichnung, tonnenförmige	BARREL DISTORTION	Distortion en barillet
Verzögerung (Phasen-)	RETARDATION	Retard
Vielstrahlinterferenz	MULTIPLE BEAM INTERFERENCE	Interférences à ondes multiples
virtuelles Bild	VIRTUAL IMAGE	Image virtuelle
visueller Nutzeffekt	LUMINOUS EFFICIENCY	Efficacité lumineuse relative
vordere Brennebene	FRONT FOCAL PLANE	Plan focal objet
Vorzeichen der Doppelbrechung	SIGN OF BIREFRINGENCE	Signe de la biréfringence
Wärmeschutzfilter	HEAT (OR HEAT-PROTECTION) FILTER	Filtre anti-calorique
Wärmeschutzglas	HEAT-ABSORBING GLASS	Verre anti-calorique
Weglänge, optische	OPTICAL THICKNESS; OPTICAL PATHLENGTH	Chemin optique
Weißstandard	WHITE BODY (OR WHITE STANDARD)	Étalon de blanc
Wellenlänge	WAVELENGTH	Longeur d'onde
Wellenlänge, kompensative	COMPLEMENTARY DOMINANT WAVELENGTH	Longeur d'onde dominante complémentaire
Wellenoptik	WAVE OPTICS	Optique ondulatoire
Wellenpaket	WAVE GROUP (WAVE PACKET)	Groupe d'onde
Wellenzahl	WAVE NUMBER	Nombre d'onde
Wellenzug	WAVE TRAIN	Train d'onde
weißes Licht	WHITE LIGHT	Lumière blanche
Weitsichtigkeit (Hypermetropie)	HYPERMETROPIA (FARSIGHTEDNESS)	Hypermétropie
Wrightsches Okular	SLOTTED EYEPIECE	Oculaire à tirette de Wright
Winkel, kritischer	CRITICAL ANGLE	Angle critique
Winkelvergrößerung	ANGULAR MAGNIFICATION	Grossissement commercial
Wollaston-Prisma	WOLLASTON PRISM	Prisme de Wollaston
Xenonlampe	XENON ARC LAMP	Lampe à xénon
Zählkammer	COUNTING CHAMBER	Cellule de comptage
Zählokular	COUNTING EYEPIECE	Oculaire pour comptage
Zeichenapparat	DRAWING APPARATUS	Appareil à dessiner
Zeichenprisma	DRAWING PRISM	Chambre claire
Zeigerokular	POINTER EYEPIECE	Oculaire à aiguille indicatrice
Zentralwellenlänge	CENTRAL WAVELENGTH	Longeur d'onde central
Zerstreuungskreis	CIRCLE OF LEAST CONFUSION	Cercle de moindre diffusion
Zerstreuungslinse	DIVERGING LENS	Lentille divergente
Zerstreungsvermögen	DISPERSIVE POWER	Pouvoir dispersif
Zentralblende	CENTRAL STOP	Obturation central
zirkular polarisiertes Licht	CIRCULARLY-POLARIZED LIGHT	Lumière polarisée circulaire
Zonenplatte	ZONE PLATE	Réseau zône
Zoom (pankratisches System)	ZOOM; PANCRATIC	Zoom; Objectif pancratique

zusammengesetztes Mikroskop	COMPOUND MICROSCOPE	Microscope (composé)
zweiachsig, optisch	OPTICALLY BIAXIAL	Biaxe
Zwischenbild, reelles	INTERMEDIARY (PRIMARY) IMAGE	Image intermédiare
Zwischenbildebene	INTERMEDIATE (PRIMARY) IMAGE PLANE	Plan de l'image intermédiaire
Zwischenlinse	INTERMEDIATE LENS	Optique de tube
Zwischentubus	INTERMEDIATE TUBE	Tube intermédiaire